HYDROLOGICAL SCIENCE AND ENGINEERING
(SERIES EDITOR - DR. TOMMY S. W. WONG)

RAINFALL FORECASTING

HYDROLOGICAL SCIENCE AND ENGINEERING

TOMMY S. W. WONG - SERIES EDITOR -

Kinematic-Wave Rainfall-Runoff Formulas
Tommy S.W. Wong (Author)
2011. ISBN: 978-1-60692-705-2

Hydrological Science and Engineering Series
Tommy S. W. Wong (Editor)

Overland Flow and Surface Runoff
Tommy S.W. Wong (Editor)
2011. ISBN: 978-1-61122-868-7

Rainfall Forecasting
Tommy S.W. Wong (Editor)
2012. ISBN: 978-1-61942-134-9

HYDROLOGICAL SCIENCE AND ENGINEERING
(SERIES EDITOR - DR. TOMMY S. W. WONG)

RAINFALL FORECASTING

TOMMY S.W. WONG
EDITOR

Nova Science Publishers, Inc.
New York

For permission to use material from this book please contact us:
Telephone 631-231-7269; Fax 631-231-8175
Web Site: http://www.novapublishers.com

Library of Congress Cataloging-in-Publication Data

Rainfall forecasting / [edited by] Tommy Wong.
p. cm.
Includes bibliographical references and index.
ISBN 978-1-61942-134-9 (hardcover)
1. Rain and rainfall. 2. Precipitation forecasting. 3. Rainfall probabilities. I. Wong, Tommy S. W. QC925.R235 2011
551.64'77--dc23

2011042527

Published by Nova Science Publishers, Inc. † New York

CONTENTS

PREFACE

Rainfall is often considered as the source of the hydrological cycle. Being able to forecast rainfall accurately has immense practical benefits. In this book, experts from around the world share their knowledge and highlight challenges on rainfall forecasting. Topics discussed include statistical downscaling for short term quantitative precipitation forecasting; empirical approaches in long-term rainfall forecasting; the challenge of warm-season convective precipitation forecasting; the precipitation field in southern Italy; prediction of tropical cyclone rainfall potential in Taiwan mountainous areas, and real time flood control using quantitative precipitation forecast and ensemble approach.

Chapter 1 - This chapter presents a review on the use of statistical downscaling techniques, their types and their application to problems related to precipitation forecast. Some results are shown from recent publications by the authors that address the use of statistical downscaling models for quantitative precipitation forecasting. It is shown that, after a careful analysis of predictors and the configuration of the statistical downscaling model, the forecast models based on statistical downscaling can be used to perform useful precipitation forecast at the six-hour lead-time. The performance of the models based on statistical downscaling is compared with the performance of a numerical weather forecast at a very high horizontal resolution. Additionally, it is shown that the use of classification techniques often used during statistical downscaling allows finding interpretations on the mechanisms acting during precipitation events. The extension of the forecasts to a probabilistic framework is also discussed by means of the comparison of results given by some frequentist and Bayesian models. These probabilistic models evolve from the best model identified in the previous quantitative deterministic precipitation forecast exercise. Despite the fact that the proposed models yield useful information on precipitation forecasts, they still fail to yield correct forecasts of intense precipitation events and the performance of the precipitation forecast must still be significantly enhanced. Some ideas for future developments are proposed at the end of the chapter.

Chapter 2 - Long-term forecasting of rainfall for periods longer than seven days over a region of high spatial and temporal variability is a challenging task. Various empirical or numerical attempts have been made to quantify this varying quantity. The effort taken across the globe in the prediction of rainfall has not attained 100% success. In this conjuncture, this chapter is an endeavour to take account of the various empirical techniques used to forecast the rainfall across the globe. The introduction of this chapter gives a description on the historical perspectives, outline of the basic forecasting techniques and the steps in forecasting. In the next section, a better understanding of the time series, some basic quantitative statistics, physical linkage and formation of indices and some common methods in the selection of predictors are described. Subsequent section deal with the various empirical methods used in rainfall forecasting; each explained with the examples from atmospheric science. The section describes both the quantitative (linear and non-linear) and probabilistic methods used to forecast the rainfall. This consists of simple regression, multiple regression, autoregressive integrated moving average (ARIMA) models, multiple power regression, Principal Component regression, Projection Pursuit Regression, Neural Network, Probabilistic forecast from ensembles and parametric model. The techniques discussed so far are used for giving forecasts at a single geographic location at different times. Other than these, the prediction at a spatial scale is common nowadays. Principal Component Analysis (PCA), Canonical Correlation analysis (CCA) and Numerical Weather Prediction (NWP) are the commonly used techniques for the forecasts at spatial fields. Excepting NWP, the empirical methods used in spatial prediction and their forecast verification methods are also discussed. The various measures used in the evaluation and correction of forecasts, the challenges and problems in issuing long-term forecasts, various biases, limitations and common practices to minimize errors are also discussed.

Chapter 3 - Deep moist convection presents numerous challenges for precipitation forecasting, as evidenced by objective skill scores for warm season quantitative precipitation forecasts (QPF) that are far lower than those for the cold season. Convective precipitation occurs on small-scales with large gradients in precipitation amounts and short lifetimes marked by high intensities, allowing even small spatial and temporal errors to result in poor objective skill scores and large negative impacts on some users of the forecasts. To make matters worse, convection is often forced by small-scale phenomena as well, such as convective outflow boundaries, soil moisture gradients, or even horizontal boundary layer rolls, all of which are rarely forecasted well by numerical weather prediction systems. Coarser grid models may use convective parameterizations which introduce large errors; even finer grid models suffer from deficiencies in microphysical and planetary boundary layer parameterizations.

Because of the many sources of error affecting convective precipitation forecasts, both ensemble techniques and postprocessing procedures are increasingly used to improve the forecasts. Although most ensembles have been found to work best when using perturbed initial and lateral boundary conditions to generate members, convective rainfall forecasting may be an exception where the use of mixed physical parameterizations and models yields an ensemble with increased spread and skill for better reliability and resolution. As with other weather parameters, rainfall forecasts based on an ensemble mean are more skillful than forecasts from any individual member; however, the ensemble mean smooths small-scale high-amplitude features too much, usually resulting in low biases for heavy amounts and high biases for low amounts. A probability-matching approach has been found to yield better skill and bias for the ensemble mean. Post-processing can help to improve these rainfall forecasts and may include bias-correction and neighborhood approaches.

Because traditional objective skill measures double penalize high resolution forecasts often used for convective forecasting and may not match impressions from human forecasters, convective rainfall forecasts are increasingly evaluated using new verification techniques such as object-based approaches. Such approaches can help modelers to understand deficiencies in the model, but the more complex output from the techniques may be more difficult for forecasters to understand.

Chapter 4 - This chapter revisits some aspects of the precipitation field, including its prediction, over Calabria peninsula, which is the southwest tip of the Italian boot. Calabria offers an interesting scenario for mesoclimatic studies in spatial and temporal rainfall variability. The region is located in the Central Mediterranean Basin, which has a strong seasonal variability that mirrors in large precipitation differences among seasons. Moreover, the complex orography of the country leads to different exposures of the territory to the rain-bearing air masses and large differences in rainfall can occur in few tens of kilometres.

After presenting the climatology of the rain field over the country and, specifically of heavy rainfall events (precipitation greater than 64 mm/day), the numerical analysis of an intense and destructive storm that occurred on 10–12 December 2003 was used to study the role of the physical factors involved. This case study was chosen because it is a prototype of heavy rainfall events occurring over the peninsula and more generally in the Mediterranean area.

The factor separation technique was applied to this storm and the case study clearly shows the effect of the three main physical factors involved in intense and destructive rainstorm occurring over Calabria: (1) the orography, (2) the sea, and (3) the synoptic scale forcing.

Results show that the Calabrian orography and the surface latent heat flux were important elements for the precipitation enhancement at the mesoscale. Humid marine air masses were advected by the synoptic flow toward the Calabrian steep coastal mountain ranges determining intense and abundant rainfall.

The analysis of the storm also reveals the crucial role of the synoptic-scale forcing, which can be accounted by considering a single key variable: the potential vorticity (PV). In the extra-tropical atmosphere, intense cyclones often develop under upper-tropospheric jet streams that are characterized by anomalously high values of the PV, and accurate specification of the PV distribution is necessary for the successful prediction of hazardous weather events.

The sensitivity of intense rainstorms to the PV streamer paves the way to a more advanced, yet simple, forecasting technique, based on a Limited Area Model Ensemble Prediction System (LEPS). The technique is particularly useful in data void regions as Calabria, where the presence of the sea makes ground based observations problematic.

A LEPS forecast, whose resolution is able to represent structures of the precipitation field not resolved by global models, generates different scenarios compatible with model error and initial/boundary conditions error. For each scenario, a pseudo water-vapour image is computed by the temperature of the isosteric surface of 75 mg/kg. Pseudo water-vapour images are then compared with the METEOSAT water vapour image, before the time of the expected rainfall in the target area, to classify the scenarios' reliability. The precipitation forecast by these scenarios can be used to refine the deterministic forecast.

The methodology was applied to the 10-12 December 2003 rainstorm. For this case it was able to discard the only scenario predicting comparatively lower rainfall, and to show the good performance of the unperturbed forecast.

Chapter 5 - Estimation of the tropical cyclone rainfall potential can be performed by adding together the respective satellite microwave derived rainfall values. However, this technique is essentially based upon a very simple cyclone shift-motion assumption and the cyclone track prediction results. Such assumptions are too simple to correspond well with the actual rainfall conditions, especially after landfall. The absence of taking into account the typhoon rainband rotation could induce significant rainfall location inconsistencies. Even more serious errors may occur if the coupling interactions with the orographic effects produced from mountainous areas are not considered.

In light of these problems, this chapter presents a method in constructing an improved tropical rainfall potential (I-TRaP) technique for mountainous regions by factoring in additional vital information. Using the original rainfall potential method as the basis, the typhoons' rotation speeds are estimated with geostationary satellite imagery and added to the new model. Furthermore, the changes in the typhoon rainfall patterns and rainfall intensity caused by the orographic effect are also considered together with the historical cases, thereby allowing the predicted rainfall patterns to be redistributed and rescaled to approximate the actual rainfall as close as possible.

Typhoon cases which made landfall in the hilly and mountainous regions of Taiwan were retrieved with satellite observations and tested to demonstrate the technique's performance. Results showed that the more detailed considerations rendered much better predictions of the typhoon rainfall potentials, both in terms of the rainfall locations and rainfall accumulation of landfalling typhoons.

Chapter 6 - This chapter discusses the feasibility of using quantitative precipitation forecasts (QPF) at an operational stage for structural and nonstructural flood control. It introduces the ensemble approach to assess forecast inaccuracy, providing valuable decision support information. Comparison of recent forecasts with real-time, in situ measurements can be used to determine forecast bias. Since underestimated or overestimated QPF might occur in areas surrounding the basin, alternatives to define the error are introduced. Then, spatial analysis is performed, not only within the targeted basin, but also within nearby areas. To assess the intrinsic uncertainty of forecasts, an ensemble method based on noise perturbation is proposed. To attempt an appropriate ensemble spread, a weighting system is developed according to an evaluation of spatial and temporal bias. The generated precipitation members are used to force distributed hydrologic models, and to obtain forecasts of reservoir inflow for structural flood control. Similarly, in nonstructural flood control mode, the precipitation ensemble can be generated for flood forecasts at the control point. Then, two types of systems can be distinguished, for supporting dam operation and early flood warning. These systems can evaluate precipitation forecast performance from previous time steps, to support proper and timely decisions. This chapter also treats study cases of operational river management in the Upper Tone Reservoir system in Japan, in which mesoscale forecast uncertainty was inserted into the dam release schedule. Additionally, the development of an early flood warning system is introduced, using global and regional forecasts in the Huong River basin of Vietnam. The implications of both cases are discussed, and recommendations for more generalized cases are highlighted.

In: Rainfall Forecasting
Editor: T.S.W. Wong, pp. 1-38
ISBN: 978-1-61942-134-9

Chapter 1

STATISTICAL DOWNSCALING FOR SHORT TERM QUANTITATIVE PRECIPITATION FORECASTING: SOME ACHIEVEMENTS AND LONG TRAIL AHEAD

***Jon Sáenz*[1], *Gabriel Ibarra-Berastegi*[2], *Agustín Ezcurra*[1] *and Alejandro Fernández-Ferrero*[3]**

[1] Dept. Applied Physics II, University of the Basque Country (UPV/EHU), Faculty of Science and Technology, Barrio Sarriena s/n, 48940-Leioa, Spain.

[2] Dept. Nuclear Engineering and Fluid Mechanics, University of the Basque Country (UPV/EHU), Technical School of Engineering, Alda. de Urquijo s/n, 48013-Bilbao, Spain.

[3] Consorcio de Aguas Bilbao Bizkaia, c/ San Vicente 8, Edif Albia, 48001-Bilbao, Spain.

ABSTRACT

This chapter presents a review on the use of statistical downscaling techniques, their types and their application to problems related to precipitation forecast. Some results are shown from recent publications by the authors that address the use of statistical downscaling models for quantitative precipitation forecasting. It is shown that, after a careful analysis of predictors and the configuration of the statistical downscaling model, the forecast models based on statistical downscaling can be used to perform useful precipitation forecast at the six-hour lead-time. The performance of the models based on statistical downscaling is compared with the performance of a numerical weather forecast at a very high horizontal resolution. Additionally, it is shown that the use of classification techniques often used during statistical downscaling allows finding interpretations on the mechanisms acting during precipitation events. The extension of the forecasts to a probabilistic framework is also discussed by means of the comparison of results given by some frequentist and Bayesian models. These probabilistic models evolve from the best model identified in the previous quantitative deterministic precipitation forecast exercise. Despite the fact that the proposed models yield useful information on precipitation forecasts, they still fail to yield correct forecasts of intense precipitation events and the performance of the precipitation forecast must still be significantly enhanced. Some ideas for future developments are proposed at the end of the chapter.

Keywords: Analogues, statistical downscaling, self organizing maps, random forests, rainfall forecast, quantitative precipitation forecast, probabilistic quantitative precipitation forecast, Bayesian model, lightning, rainfall yield, empirical orthogonal functions, canonical correlation analysis.

INTRODUCTION

The production of good quantitative precipitation forecasts (QPF) even with a short lead-time of a few minutes or some hours is a major objective of current meteorology. Extreme precipitation events pose great threats to human lives and huge economic losses. Therefore, meteorological services providing precipitation forecasts of the best quality is a need of emergency managers and decision makers. Additionally, producing accurate QPF, particularly for heavy rainfall events is still a scientific challenge, with several authors exploring different techniques in order to achieve that goal (Ebert et al., 2004; Fox and Wikle, 2005; Shao, 1998; Smith and Austin, 2000; Xiao and Sun, 2007; Xu et al., 2006). In general, advanced QPF systems use several sources of information, such as numerical weather forecast (NWF) models, statistical techniques, radar or observed rain-gauge data and satellite images.

Since the times during the mid XXth Century when forecasts by NWF models were for the first time better than the ones produced by human forecasters (Charney et al., 1950), there has been a steady increase in the use of NWF model output for weather prediction. However, even when run at the highest spatial resolutions than can be used with today supercomputers, there are several aspects of the processes producing rainfall that are not properly resolved by the NWF at the grid scale. On the one side, the observations leading to analysis used for the initial condition are in some cases too sparse or noisy and several features of the atmosphere (clouds, fluxes through narrow valleys or localized moisture plumes) cannot be properly identified. On the other side, precipitation systems are very often quite localized in space and time, and the resolution of the grid of the NWF does not allow sampling or forecasting them adequately, particularly for convective precipitation events. Finally, the NWF model is just an approximation to the real physical processes acting during the development of the precipitation systems. Some of the assumptions done during the formulation of the NWF model, such as parameterizations of radiation, clouds, microphysics or surface turbulent fluxes are just reasonable approximations to the real processes. Therefore, it is convenient to complement information derived from NWF by other means in order to achieve a precipitation forecast of good quality (Bauman et al., 1997; Golding, 1998; Kidson and Thompson, 1998; Lack and Fox, 2007; Xu et al., 2006; Yates et al., 2006; Vasiloff et al., 2007).

Several groups have developed methods based on advection routines over observed radar data (Boudevillain et al., 2006; Fox and Wikle, 2005; Thielen et al., 2000), and these methods yield useful forecasts for short lead times, since radar data do not generally allow to extrapolate the future development of a given precipitation event. The estimation of errors for this kind of radar based QPF is an active research field (Kirstetter et al., 2010).

There is an alternative path, which is to identify useful statistical relationships between the NWF output and corresponding observations during a training period. These statistical relationships allow calibrating the model output so that the NWF QPF better fits observations (Altava-Ortiz et al., 2006; Benestad et al., 2008; Fernández-Ferrero et al., 2009; Hart et al., 2004; Kotroni and Lagouvardos, 2004; Obled et al., 2002; Yuval and Hsieh, 2003, Wilby et al., 2004; Wilby and Wigley, 1997). In this case, the set of statistical techniques used to improve the quality of the NWF forecast includes multiple linear regression models, Kalman filters, classification techniques, neural networks or analogues, amongst others.

The idea behind analogue-based methods is that, given a sufficiently long database of past observations of precipitations at one or several observatories together with the analysis or NWF outputs corresponding to those days, the best forecast for the current day that can be issued is the one corresponding to the closest synoptic state in the database of observations (Obled et al., 2002;

Zorita and von Storch, 1999). In general, as the information in the predictor fields is quite redundant, the analogues are usually searched for in a space of reduced dimensionality, most of the times the one spanned by the empirical orthogonal functions (EOFs) but it can also be the space spanned by the leading canonical correlation patterns (Fernández and Sáenz, 2003) if the field to be predicted needs to be involved in the forecast.

Neural networks are very often used for short-term prediction of meteorological variables due to their ability to model non-linear mechanisms (Ibarra- Berastegi et al., 2008). Neural networks have been applied to forecasts of short-range precipitation up to six hours in advance, wind damage or thunderstorms (Kuligowski and Barros, 1998; McCann, 1992; Marzban and Stumpf, 1998).

Self Organizing Map (SOM) consists in a group of artificial neural networks based on the topological properties of the human brain. Readers can refer to Kohonen (1982), Haykin (1994) and Patterson (1996) for fundamental details on the technique. Self-Organizing Maps have found application in meteorological and climate studies to detect anomalies in the synoptic circulation patterns, extreme events or forecasting purposes (Cavazos, 1999; Cavazos 2000; Gutiérrez et al., 2005; Hewitson and Crane, 2002; Sheridan and Cameron, 2011). SOMs are also known as Kohonen networks in honour of the author who first developed them. SOMs are primarily intended to identify groups or detect patterns in a set of data. They are used with that objective in mind as shown by some results presented in this chapter.

This chapter shows several results of the application of statistical techniques to NWF model output over the Basque Country, northern coast of the Iberian Peninsula (Figure 1). This area shows an abrupt topography, with a steep meridional gradient of precipitation and temperature, gradient due to the combined effects of continentality and topography (Font-Tullot, 2000). The data in this chapter cover places close to the coast such as Bilbao (BI) and San Sebastián-Donostia (SS) and another one Vitoria-Gasteiz (VI) of a more continental origin when the analysis or rainfall yield and lightning is performed. For the analysis of statistical downscaling in short-range QPF, the observatories and the topography of the area are shown in Figure 1. Finally, the application of Random Forest to compute daily rainfall amount in observatories in the Ebro Valley covers Zaragoza (Z) and Tortosa (TO).

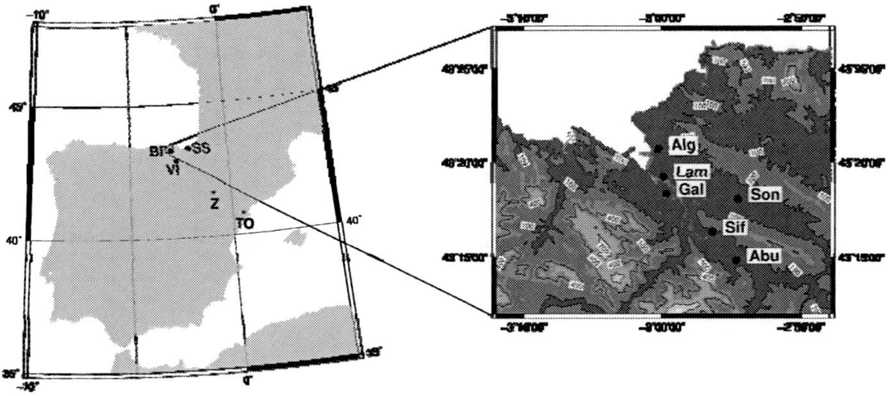

Figure 1. The square in the map shows the area where the data used in this chapter for six-hourly QPF have been recorded. The inset shows a detailed topography of the smallest domain used in some of the parts of this study. Data on lightning and rainfall have been recorded at the circles (VI, BI and SS). Precipitation data for the application of analogue methods plus multiple linear regression or random forests comes from the places marked with a star (Z and TO).

The main objective of this chapter is to briefly show the results of the application of several classification techniques and other statistical downscaling methods to outputs from numerical models, with the aim both to diagnose and forecast precipitation processes over the study area. The precipitation forecast systems covered in detail in this chapter are mainly focused on short-range QPF. In order to properly cover this objective, the results of several sensitivity tests to predictors, domains or configuration of the models will be shown, together with a comparison of the results of these models with other results from the literature or other alternative precipitation forecast systems developed by our group. Additionally, those results will be analyzed from the point of view of the physics of precipitation and lightning systems, in order to properly diagnose the results.

DATA

In the analysis of the connection between lightning count and total daily precipitation, a database was built with 429 cases. Each case represented one day when an electrical storm was recorded. Every case was associated to a point in a 14-dimensional hyperspace in which each axis represented a particular variable from the following list:

- Daily cloud-ground (CG) lightning counts detected by the Spanish National Meteorological Institute direction finder network in a window of 20,800 km^2 covering the region shown in Figure 1, located between coordinates 42° 24’ to 43° 36’ N and 3° 36’ to 1° 24’ W. All cases in that window in which at least 1 CG was detected were retained.
- Daily CG counts detected over three windows of 20x20 km2 centred over the three largest metropolitan areas in the region shown in Figure 1: Bilbao (BI), San Sebastián (SS) and Vitoria (VI).
- Daily total rainfall depth collected over the above-mentioned urban areas.
- Daily mean temperatures at 850 and 500 hPa.
- Airflow directions observed at 850 and 500 hPa, represented in every level and case by the two wind (U and V) components associated to the eight directions in the wind rose. In the situations in which the observed pressure gradient was too small, these values were assigned zero wind speed for the two components.
- The maximum surface temperature reported for the area.

The SOM was used to cluster observations of all these variables within a reduced number of groups in which rain yield and synoptic situations were expected to be similar.

The results showing the performance of QPF models derived from statistical downscaling are based on the analysis of observations recorded by six rain gauges Galindo (Gal), Sifón (Sif), Abusu (Abu), Algorta (Alg), Lamiako (Lam) and Sondika (Son) placed over the Bilbao sewage network. The data were available from September 2000 to December 2003 with 10’ temporal resolution, which was averaged to 6 h. The other example mentioned in this chapter regarding precipitation forecasts involves precipitation at two different observatories (Tortosa, TO and Zaragoza, Z) placed at the coast and in the interior of the Iberian Peninsula, in the Ebro Valley. In this case, the data used correspond to daily precipitation values covering the period 1961-2001 and were obtained from the ECA repository (Klein Tank et al., 2002).

Data on the atmospheric fields used in all the parts of this chapter (lightning and rainfall, and precipitation forecasts) have been obtained in all the cases from the European Centre for Medium-Range Weather Forecasts’ (ECMWF) Meteorological Archive and Retrieval System (MARS) server and correspond to data from ECMWF’s 40 Years Re-Analysis (ERA40) (Uppala et al., 2005), ERA Interim and operational archives, depending on the period used in each case (ERA40 finishes in August, 2002, whilst ERA Interim starts in 1989). In all the cases, the atmospheric analyses have been downloaded with a horizontal resolution of 2.5° for ERA40 and 1.125° for ERA Interim and operational analysis. The downloaded fields comprise several variables at different pressure levels (geopotential, specific humidity, zonal and meridional components of wind and temperature) plus additional fields at the surface, such as mean sea level pressure, temperature and dew point temperature at 2m, wind speed components at 10 m, skin temperature of the ocean, snow cover depth or soil temperature. The dew point temperature has been computed from the specific humidity by means of an approximate formula derived from mixing ratio and pressure (Bohren and Albrecht, 1998). These data have been used to provide predictor fields to the different statistical downscaling models used or to compute additional diagnostic fields such as convective instability indices, vertical wind shear through the troposphere or vertically integrated moisture transports, amongst others. The instability indices comprise the K index (K=t850-t500+td850-t700+td700), the Showalter index $S=t_{500}-t_L$ and the Totals index (TI=td850+t850-2 * t500), where t850, t700 and t500 correspond to temperature (expressed as degrees Celsius) at 850, 700 and 500 hPa isobaric surfaces; td850 and td700 indicate dew point temperature (degrees Celsius) at the 850 and 700 hPa isobaric surfaces and t_L represents the temperature (degrees Celsius) of a parcel lifted by means of a dry-adiabatic process from 850 hPa to the lifting condensation level and moist adiabatically from the lifting condensation level to 500 hPa. The indices quantify the convective instability of the atmospheric column during the storm days. Values of K index above 24°C, very low values of S index and TI above 44 indicate likely thunderstorms (Djurić, 1994). Vertical wind shear is computed by means of the difference (modulus) of wind speed between 700 and 300 hPa. The evaluation of vertical integrals for the computation of moisture transport has been performed in the original model levels using the original spectral representation, thus precluding the introduction of spurious errors during vertical or horizontal interpolation (Trenberth and Guillemot, 1995).

Additionally, the coarse-resolution data from the ECMWF's archive has been used to force a run with three nested domains (D1 at 81 km horizontal resolution, D2 at 27 and D3 at 9 km horizontal resolution) of the MM5 mesoscale model (Grell et al., 1994) without assimilating observations during the integration of the MM5 model. Further details on the configuration of the PSU/NCAR mesoscale model (known as MM5) can be obtained from Fernández-Ferrero et al. (2009). The same predictors used from the global reanalysis and mentioned above have also been used from the MM5 model output at the three D1, D2 and D3 domains.

METHODOLOGY

Classification Techniques (SOM) Applied to Study the Relationship between Lightning and Rainfall

Self Organizing Maps (SOMs) are also known as Kohonen networks in honour to the author who first developed them. SOMs are primarily intended to identify groups or detect patterns in a set of data. This statistical tool is designed for unsupervised learning and is trained using the Kohonen algorithm. This algorithm does not require any examples of the output variable in the data set. The nodes of the output layer are arranged on a topological map, forming a two dimensional lattice. The Kohonen training algorithm is designed to encourage the formation of clusters of similar cases at nearby positions in the lattice. This allows independent clusters to be identified and labelled, even in the absence of labelled training data. Class labels are assigned to nodes, and it is the class label of the "winning" node during execution that forms the class or grouping estimation.

SOM can also be understood as a non-linear mapping of a multidimensional hyperspace onto a two dimensional plane. Points which are close to each other in the hyperspace can also be expected to be close to each other (i.e. assigned to the same node) in the two dimensional lattice. Accordingly, clusters of points or patterns can be identified because they are usually assigned to the same node.

Besides using SOM to find the best combination of data for the training, validation and test datasets used to train the neural network, in this chapter we also show the results obtained using this classification technique to diagnose the relationship between lightning and precipitation. The final objective of applying SOM method in this case was to discover meteorological patterns associated to the lightning and rain production in the Basque Country area. For that analysis, daily cloud-ground lightning counts and daily-accumulated precipitation were first analysed by means of principal component analysis in order to isolate the main variability of the combined fields. The same database used in PCA has been employed during the SOM analysis but adding a new group of meteorological data deduced from synoptic maps given by INM for the related days. These new group of variables were: temperature and air flow direction observed at 850 and 500 hPa (clustered in the 8 rose wind directions and one more labelled L – local class –when the geostrophic wind module was expected to be near null). The maximum surface temperature reported for the area was also taken into account.

The SOM analysis was carried out with an input layer of 14 nodes and an initial 3x1 structure (3 nodes) in the output layer with the aim to confirm or not the clustering results suggested by previous studies (Ezcurra et al. 2002). The whole SOM analysis was designed in an interactive way so that the final number of nodes, i.e. the final dimension of the output layer, would be identified only if the cases assigned to each node were similar enough as to constitute a pattern.

SOM was fitted, as usual, in two phases. In the first one, 100 epochs were used with a changing learning rate ranging from 0.1 at the beginning to 0.02 at the end. The neighbourhood parameter started from a value of 3 and finished with a value of 1. In the second phase, 1000 epochs were used with a learning rate ranging from 0.1 to 0.01 at the end. In this phase, the neighbourhood parameter had a constant value of 0, corresponding to the standard Kohonen "winner takes all" algorithm. The 3×1 SOM was built using the standard "winner takes all" algorithm A SOM with this 3×1 structure (3 nodes) in its output layer yielded meaningful and consistent results, thus grouping similar cases in the same node. This 3×1 structure could be understood as the most proper projection of all the events onto an easy-to-handle two-dimensional representation.

Deterministic Statistical Downscaling by Means of Analogues

As usual in statistical downscaling exercises, the main goal of the studies described in this chapter was to develop a transfer function between the atmospheric state $x(\vec{r},t)$ at the large scale explicitly resolved by the NWF models performing the atmospheric analysis or weather forecasts and the observations (precipitation, lightning count, wind or the variable that is going to be predicted) at the observatory scale $y(\vec{r}\,',t)$, with a primed $\vec{r}\,'$ denoting space in this last case, since the domain of the predictors and the predictands (variable to be forecast at the local scale) will normally not be the same. Literature on statistical downscaling methods shows that there exist several options in the development of downscaling exercises. There exist univariate or multivariate linear regression methods, classification methods (based on clustering techniques or other techniques such as self-organizing maps) or other nonlinear methods such as the analogue method or neural networks. In some cases, those methods are applied to the coarse resolution model outputs and in other cases they are applied to high-resolution numerical models nested in the output of the coarse resolution models. There exist several reviews on the topic and the reader is referred to them for a thorough discussion (Benestad et al., 2008; Fernández-Ferrero et al., 2009; Fernández-Ferrero et al., 2010; Weichert and Bürger, 1998; Wilby and Wigley, 1997). In the following pages, examples will be presented on the performance of different methods based on analogues, neural networks, classification by means of SOM and random forests, so that the focus in this section will centre on those techniques.

In the case of analogues, the basic hypothesis is very simple. Given a state $x(\vec{r},t_0)$ of the atmosphere corresponding to the day that the forecast must be issued, we expect that precipitation (or any other variable observed at the site scale) will appear in a quantity similar to the amount $y(\vec{r}',t_i)$ recorded when the observed atmospheric state $x(\vec{r},t_i)$ was the closest one to the observed at the day the forecast is being issued (Wetterhall et al., 2005; Zorita and von Storch, 1999). The analogue method was initially used for weather forecast, but it was severely criticized because of the difficulty of finding good analogues (Lorenz, 1969; van den Dool, 1994; Nicolis, 1998). This problem is easier to overcome at a regional scale (Roebber and Bossart 1998) and it is not a serious problem for short lead times (Hamill and Whitaker 2006; Stephenson et al., 2005).

In general, given that the atmospheric state is described by means of a set of numbers in the grid of the NWF (or climate) model, the numbers of the grid show a high spatial correlation for most of the variables. Therefore, it is customary to initially compress the information given by the predictors by means of an EOF analysis, reducing the dimensionality of the space where the search of similar synoptic states is to be performed. If the predicted variables also share a high fraction of variance, the EOF analysis can also be performed on the predicted fields, too, since this can filter some observational error. During the computation of the EOFs, several options exist.

- If two or more predictors are used (say sea level pressure and temperature) and their variances are very different, the leading EOFs would tend to overestimate the role of one of the variables if combined EOFs (Bretherton et al., 1992) are used. Therefore, if the variances of the different variables are not of the same order of magnitude, the EOFs are usually computed on the correlation matrix or by means of the singular value decomposition of the standardized data, depending on the positive definiteness of the correlation matrix. This depends on the sample size and the number of grid points to be used in the forecast.

- An alternative strategy might be to compute the EOFs corresponding to each of the fields and then, find the analogues in the hyperspace obtained by combining all the principal components obtained by means of the EOF analysis.
- Additionally, it is also customary, depending on the design of the downscaling model, to remove additional components of the input data such as the seasonal cycle (the forecast might be produced by means of anomalies or by means of centred data) or even the linear trend. This has to be decided at each case, depending on factors such as the seasonality in the data or the presence or absence of a linear trend component in the data.

The problem with the approach of finding the analogues in the subspace spanned by the leading EOFs is that the same set of predictors would yield the same set of analogues for different variables such as for temperature or precipitation forecasts. However, it is well known that different physical mechanisms affect the behaviour of temperature or precipitation. In order to involve the quantity to be predicted (predictand) in the space used for the identification of analogues, the search can be done in the space of the canonical correlation coefficients. This showed a better skill with lower dimensionalities of the search space (Fernández and Sáenz, 2003). The use of a subspace with a smaller dimension is a good option when the sample size is not very long. However, this can be shown to be equivalent to a solid rotation in the space of the EOFs if no additional truncation is used after the EOF computation before conducting the computation of the canonical correlation coefficients (Fernández and Sáenz, 2003).

Once the predictor fields (atmospheric synoptic conditions) $x(\vec{r},t)$ have been compressed to the EOF space forming a series of reduced dimensionality $p(\vec{s},t)$, the difference between two atmospheric states $p(\vec{s},t_0)$ and $p(\vec{s},t_i)$ is usually measured by means of a norm $d_{0i} = \left\| p(\vec{s},t_0) - p(\vec{s},t_i) \right\|$ in the subspace spanned by the leading EOFs, and several options exist for the selection of the norm. The use of the euclidean norm or the cosine of the angle between the two vectors in the EOF space, are common selections for norm, but others exist (Matulla et al., 2008; Toth, 1991). After the distances d_{0i} in the EOF space are computed by means of the selected norm, they are sorted and the smallest one (or the smallest ones) becomes the analogue (or set or analogues) on which the next steps of the algorithm are performed. In some cases, anti-analogues (the biggest d_{0i} can be used to produce the opposite anomaly in the output field) are also used.

The simplest next step is to use the values $y(\dot{r}',t_i)$ as the forecast for day t_0. An alternative approach can be to use the average of the closest ***N*** analogues as the predicted value at time t_0. In general, as the user increases the value of ***N***, the correlation between forecast and observed values of $y(\dot{r}',t)$ is better, but to the expense of the variance of the series (Fernández and Sáenz, 2003).

In other cases, the search of analogues is just an initial phase of the downscaling exercise. After the selection of the analogues, a canonical correlation analysis on the subset of synoptic situations found close to the day being forecast can be used, as shown in the results presented from Fernández-Ferrero et al., (2009). In this chapter, the performance of a multiple linear regression model and random forests applied after the initial selection of a set of analogues will also be shown. This design in two stages after the initial selection of a subset of synoptic situations followed by a second step is common in the literature (Gutiérrez et al., 2004).

In the results shown in this chapter, several downscaling models have been used, but only the main results will be shown, and further information can be found in the original papers, which, for the shake of the description of the sensitivity to the predictor is (Fernández-Ferrero et al., 2009). As predictors, geopotential height, zonal and meridional components of wind velocity, temperature and dew point temperature from ERA40 and the output from an integration of the MM5 using the same ERA40 data as initial and boundary conditions have been used. The results of the precipitation simulated by the MM5 model have also been used. In our case, the principal component analysis has been performed on the predictor and precipitation data. After the EOF computation, a set of the best ***N*** analogues, as measured by the euclidean distance in the EOF space are used for the next phase. Next, a classical canonical correlation analysis (CCA) is used with the data corresponding to the best ***N*** analogues, so that for each forecast time, the axis where the CCA is used change, depending on the best analogues found for that time step. Due to the combined use of analogues plus CCA, the results from this technique will be denoted as AnCCA onwards. Further details of the algorithm such as the variables used in every model, the truncation points during the computation of EOFs or the values of N can be found in Fernández-Ferrero et al. (2009).

Deterministic Statistical Downscaling by Means of Neural Networks

For the same data mentioned above, a deterministic precipitation forecast model has also been formulated using neural networks (NN). In general terms, an NN can be understood as a nonlinear function that maps predictors (inputs) into precipitation, wind or the variable to be predicted (outputs). The mapping function is derived during a training phase after selecting the network architecture. The definition of the architecture of the NN means that the user must choose the number of layers, the number of nodes per layer and the number of connections. The NN learns how to optimally associate inputs and outputs (Masters, 1993; Bishop, 1995; Ibarra-Berastegi et al., 2008) by means of a mapping function. After the network architecture is defined, the values of the parameters corresponding to the mapping function are fitted generally starting from random values.

The Multilayer Perceptron (MLP) is the most popular type of NN because a MLP network with one single intermediate layer and a sigmoid activation function is in theory able to approach any function after a proper training process (Masters, 1993). If the relationship between the variables to be mapped by the MLP involves a discontinuous function, two hidden layers may be required (Masters, 1993). That is the reason why in general more than one hidden layer is used. It provides greater flexibility and enables approximation of complex functions with fewer connection weights (Masters, 1993). The main features of MLPs and NNs can be found in the literature (Masters, 1993; Bishop, 1995).

In this chapter, NNs were used besides the AnCCA technique in order to intercompare their results. The best of the AnCCA models was chosen and its performance was compared with the best NN obtained using the same set of inputs and outputs. The outputs of those models are the principal components used to reconstruct the average precipitation during the forthcoming six hours at the six rain gauges. Therefore, three different sets of NNs were built, each with 21 candidate inputs and one output. The single output of every neural network corresponds to each of the three principal components explaining most of the variance of the six rain gauges. Precipitation at the six rain gauges is reconstructed from these three principal components given as output by the neural networks.

In general, statistical models try to learn from the past in order to forecast similar future situations. Thus, the available data is split into two independent parts. The first one is a development set that is only used to build the model. The rest of the sample remains as an independent test set. However, this general rule does not apply exactly to NNs because over-fitting must be avoided (Masters, 1993; Bishop, 1995). Otherwise, under over-fitting conditions, the NNs yield inaccurate predictions for new observations. The problem of over-fitting can be avoided by using a third independent validation data set to assess the performance of the model at various stages of learning. Training will stop when the error in this validation set reaches a minimum. This approach is known as early stopping. Thus, for using early stopping the available data need to be divided into three subsets: training, validation and testing (Masters, 1993; Bishop, 1995).

Even though over-fitting may be avoided by means of early stopping, there is another problem to be avoided in order to achieve a good generalization performance. The NN must be faced with data from the test set that are not very different from the examples that it saw during training and validation. That is, the statistical properties of the data used to train, validate and test the network must be similar. The way that available data are divided into training, validation and test sets is a critical issue as shown in the literature (Maier and Dandy, 2000; Bowden etal., 2001; Maier et al., 2004). Appropriate techniques to divide data are genetic algorithms and Self-Organizing Maps (SOMs) (Maier and Dandy, 2000; Bowden et al., 2001; Maier et al., 2004).

Once data is split into training, validation and test sets, the network fitting process starts, usually from random values of the parameters of the network that are going to be estimated. Sometimes, this process may get stuck in local minima. The selection of an appropriate convergence algorithm may avoid this and help reach the minimum absolute error. At this point, the best values for the network parameters are obtained. The selection of the most suitable convergence algorithm can make it easier to reach the minimum absolute error, obtain the optimal values of the network parameters and, as a consequence, improve the generalization performance of a given NN, but only to a certain extent (Lu et al., 2002). In the case of MLP networks, Back Propagation (BP) is generally used during the initial stages of convergence, followed by the Conjugate Gradient Descent (CGD) algorithm at the end (Masters, 1993; Bishop, 1995).

For the forecast of precipitation over Bilbao by means of statistical downscaling, 5233 cases with 24 variables (21 inputs+3 outputs) each were available. These cases had to be divided into training, validation and test sets and we used SOMs for that purpose. The final number of cases in which the original data set was divided included 3134 for training, 530 for validation and 1569 for testing.

Use of Random Forest for Daily Rainfall Forecast

In the last decade, a great number of machine learning algorithms have been developed (Witten and Frank, 2005). Among them, Random Forests (RF) has become increasingly popular for several reasons, being the most important one its ability to model nonlinear relationships.

RF is based on the CART technique, in which a tree is built to relate a set of inputs to a group of output variables. The relationship between inputs and outputs may be highly nonlinear if taken as a whole, but if analyzed at different ranges, the nature of the relationships inside each range may be modelled in a simpler way. For this reason, in a CART the feature space is recursively splitted into a set of regions in which a simple model like a constant (Hastie et al., 2001) or a simple linear regression can be fitted. As the splitting process progresses starting from the root, the tree is divided at subsequent nodes into branches and sub-branches until reaching the leafs (a more homogeneous region where a simple regression between inputs and outputs is more evident). The final stage implies going backwards to the tree obtained and prune it using cost-complexity pruning algorithms. (Hastie et al., 2001).

Random forests are built using CART but adding two layers of randomness (Liaw and Wiener, 2002):

1. If a number of bootstrap samples from original data are drawn, a regression tree can be obtained with each sample, thus obtaining a 'forest'.
2. In standard trees, each node is split using the best split among all variables. In a random forest, for each tree, each node is split using the best amongst a subset of m predictors randomly chosen at that node.

The only relevant parameter in running the RF algorithm is m (the number of predictors made available to each node). Being p the number of candidate predictors, values of m like p/3 or $p^{0.5}$ are usually adopted. Nevertheless, results do not strongly rely on the choice of m (Breiman, 2001).

RF can be used for classification and regression purposes. In the case of regression, the output of the RF model will simply be the average of the outputs obtained in each tree. RF is reported to clearly outperform CART (Siroky, 2009), being one of RF's most important advantages when compared with other techniques, that over-fitting never takes place (Siroky, 2009). The relative importance of each predictor on the predictand is estimated at each tree by calculating the increase in the mean square error due to permuting a given predictor. If that variable has no predictive value for the response, it should not make a difference if its values are randomly permuted before the predictions are generated (Grömping, 2009). Then, differences due to these permutations in all the trees are averaged and normalized by the standard error. This is known as mean decrease in accuracy of the predictor.

In one of the examples shown below, analogues followed by RF have been used for estimating daily precipitation through downscaling (Ibarra-Berastegi et al., 2011). The results have been compared with the plain use of analogues and also, analogues followed by multiple linear regression. In all the cases, models have been fitted with ERA40 data and results are provided using ERA40 and also ERAInterim data to feed the models. The use of a different set of analysis (ERA Interim) yields an estimation of the sensitivity of the results on the bias between the training (ERA40) and application (ERA Interim) datasets. This estimation will be limited by the fact that this bias is probably low, if compared with other cases that might appear during operational application of this method or for climate change scenarios.

Probabilistic Rainfall Forecasts Using Analogues

The computational simplicity and performance of methods based in the search of analogues make them a good option to develop probabilistic rainfall forecasts either for ensemble-based seasonal forecasts (Feddersen and Andersen, 2005) or for operational ensemble forecasting (Diomede et al., 2008; Hamill and Whitaker, 2006). The advantage of using a probabilistic precipitation forecast is that the precipitation forecast is linked to a quantification of the uncertainty, thus allowing the end user to make a better use of the forecast (Diomede et al., 2008; Kruizinga and Murphy 1983; Krzysztofowicz 2001).

There exist several systems in the literature that yield a probabilistic precipitation forecast. For instance, Gutiérrez et al. (2004) developed a system based in the preselection of a set of m-means clusters followed by kernel based probability density function estimation around the cluster's centroids. In other cases, the downscaling method initially yields some statistical moments followed by weather generators (Busuioc and von Storch, 2003; Marty et al., 2008). Obled et al. (2002) proposed a system based on analogues in which after an initial stage of analogue search, a parametric fit of an incomplete gamma function was used to yield probabilistic precipitation forecast. The use of a preselection of analogues as the initial step in the generation of a probabilistic precipitation forecast is very common in the literature (Diomede et al., 2008; Hamill et al., 2004; Hamill and Whitaker 2006; Kruizinga and Murphy 1983), with different ways to produce the probabilistic forecast after this initial selection. In the last paper that we survey in this chapter (Fernández-Ferrero et al., 2010), several frequentist and Bayesian methods are used to yield a probabilistic forecast after the preselection of a set of analogues in the first phase of the algorithm.

In order to develop a probabilistic precipitation forecast, several precipitation thresholds $\boldsymbol{u_i}$, with $\boldsymbol{i = 1 \ldots N_u}$ are defined, with $\boldsymbol{u_i}^*$, the middle point of the interval $\boldsymbol{(u_i, u_{i+1})}$. In the first case, the probabilistic QPF model (called A onwards) is probably the simplest one that can be built. Given the set of preselected analogues, they are used to compute the probability of exceedance for the precipitation thresholds just by counting the amount of analogues, which would produce precipitation amounts higher than the thresholds. The other three probabilistic models involve the use of Bayes' theorem. The posterior probability that the precipitation at a rain gauge corresponds to each of the intervals is given by:

$$P(y = u_i^* | \mathrm{D}_i) = \frac{P(D_i | u_i^*) P(u_i^*)}{\sum_{j=1}^{N_u} P(D_j | u_j^*) P(u_j^*)} \tag{1}$$

with $\boldsymbol{D_i}$ the average distances of all the analogues yielding precipitation belonging to class $\boldsymbol{u_i}^*$ and the likelihood $P(D_i | u_i^*)$ the probability of getting a distance $\boldsymbol{D_i}$ if observed precipitation was $\boldsymbol{u_i}^*$. The different Bayesian models differ in the way the likelihood is computed. The first Bayesian model B uses as prior the frequentist estimation of the probability that is found by the model A and the likelihood is parametrically fitted to observations on the basis of the distance between analogues using as distance the states of the atmosphere, without involving precipitation. In the next two cases, in order to yield a better discriminating power of the likelihoods, the rainfall is included in the estimation of $\boldsymbol{D}_i$ but, since it is not known at the moment of the forecast, it is numerically estimated by minimizing the squared error computed from the posterior probability. Full details of the algorithm can be found in Fernández-Ferrero et al. (2010). Bayesian Experimental (BE) uses as likelihood a normal distribution (log-normal for low precipitation amounts) fitted to the observations. The last Bayesian model, Bayesian Idealized (BI) uses a likelihood that follows an idealized potential form in order to check the sensitivity of the algorithm to the form of the likelihood.

Results

Lightning and Rainfall

Results of the SOM analysis assigned 111 cases (25.87%) to the node 1 (N1), 150 (34.97%) to the node 2 (N2) and 168 (39.16%) to node 3 (N3). The analysis of the seasonality of the distribution of cases shows that those ones classified as N1 are very frequent through the year, except during summer months, when they are absent. N2 appears through the year, including summer months, but it is most frequent during spring and autumn. Finally, cases classified as N3 only appear during summer months and are very infrequent (though some scarce cases appear) during late spring or early autumn.

In order to identify the synoptic conditions that are the most frequent during the cases classified as N1 to N3, Figure 2 shows the geopotential height anomaly (seasonal cycle removed on a monthly basis from daily geopotential height data), the K instability index and the zonal and meridional components of the vertically integrated moisture fluxes over the domain. These meteorological maps were plotted using ERA 40 data associated to every case found in each node. Finally, Table 1 resumes rain yield characteristics associated to the found nodes.

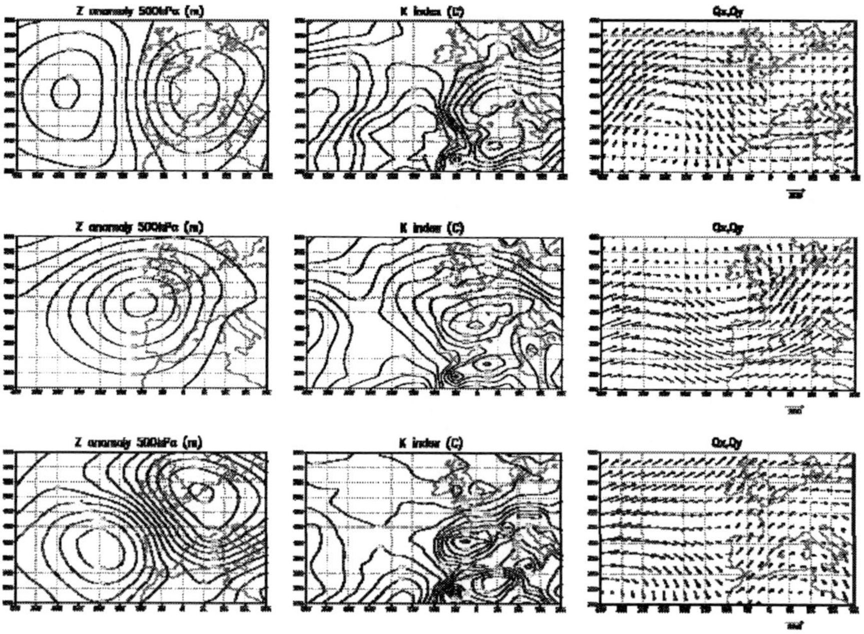

Figure 2. From top to bottom, circulation anomalies corresponding to lightning events classified by the SOM analysis in the first cell (top row), the second one (middle) and the third one (bottom row). From left to right, geopotential height anomaly of the 500 hPa surface, K instability index corresponding to the days classified at each cell and vectors representing zonal and meridional components of the vertically integrated moisture transport through the troposphere.

The synoptic situation corresponding to the storm events conforming the first node of the SOM analysis (N1) is characterized by a wave-like 500-hPa geopotential height field corresponding to Oceanic regimes with large-scale advection of humid air from the Gulf of Biscay. K instability index values reach extremes of 15°C as shown in Figure 2 (TI values reach maximums of about 45). The associated instability affects a large area centred over the Basque Country region. The cloud-ground flashes are more frequent in the coastal zone, probably indicating that topography is playing a significant role in this case. Mean rainfall yield findings (expressed as 10^4 m^3 of rain per CG count flash) range from 267 at Bilbao to 1387 at Vitoria, exhibiting an important spatial variation, consistent with the role played by topography (Table 1).

Days classified under N2 yield a substantially different meteorological structure if compared to the previous case. The geopotential height anomalies at 500 hPa (Figure 2) is unimodal with maximum negative values of -120 m, linked to a thermal anomaly of about -4 °C at 500 hPa, localized west of the Basque Country area. The events clustered into node 2 present advection of humid air flowing from the south and west of the Iberian Peninsula, thereby associated to continental regimes. The instability indices K display values of about 24°C, as shown in Figure 2, with TI values reaching extremes around 48. The associated instability covers a more extended area than observed in the previous case. These events are clearly linked to cold thermal anomalies placed at high levels in the troposphere. Mean rainfall yields in N2 (expressed as 10^4 m^3 of rain per CG count in Table 1) range from 123 at Bilbao to 183 at Vitoria. Again we found some spatial non-homogeneous distribution of rainfall yield although the spatial variation in this group is less intense.

The third SOM node (N3) is connected to events characterized by a wave-like geopotential height field at 500 hPa almost in quadrature, yet weaker when compared to the N1 cases. Geopotential height anomalies at 500 hPa reach maximum values of +40 m (see Figure 2) placed over northern Europe. Total advection of moisture arises from the south or Iberian Peninsula but with a near zero geostrophic wind at 850 hPa over the area. The instability indices show (K index shown in Figure 2) very localized high values of about 24°C over land areas, with TI values around 48, indicating that the instability of this type of meteorological pattern is mostly linked to the diurnal shooting of convection from hot land areas (local instabilities). In this cluster we found the lowest values for Mean Rainfall Yield (expressed as 10^4 m^3 of rain per CG count in Table 1) ranging from 10 at Bilbao to 24 at San Sebastián, due to the fact that this meteorological pattern produces scarcer rain although lightning production remains considerable.

These findings clearly suggest the existence of three different meteorological patterns producing a range of rainfall yield values. Those patterns agree with the previous ones established by Ezcurra et al. (2002) and the average rainfall yield is consistent with the thesis of the above-mentioned paper.

Table 1. Values of Mean Rainfall Yield (MRY) expressed in 10^4 m^3 of rain per CG count for every cluster identified through SOM analysis

NODE	MRY-BI	MRY-SS	MRY-VI
N1	267	416	1387
N2	123	141	183
N3	10	24	13

The SOM results pointed to a strong meteorological influence on rain and lightning production appearing in the area. One type of electrical episodes (N1 type) is formed under oceanic advection (NW air fluxes) created by baroclinic perturbations. In such cases mean rainfall yield estimates reach values of 700 10^4 m^3 per CG count. The rainfall yield presents a strong spatial variation from coastal places to the interior one.

The second type of electrical storms (N2) is associated to local instability shooting by surface heating with advection of humidity flowing from the Iberian Peninsula. During such meteorological events, the minimum rainfall yield registers a mean value of 15 10^4 m^3 per CG count.

The South West (SW) air fluxes associated with cold air in upper parts of the atmosphere (N3) represents the final meteorological cluster. In those cases mean rainfall yield in the area is about 150 10^4 m^3 per CG count.

In all electrical episodes K instability indices is greater than 15°C but in the most productive lightning episodes this index may at times reach values greater than 24°C.

Deterministic QPF at Short Range (6 Hours)

For the deterministic forecast of precipitation, there are several questions that were addressed by Fernández-Ferrero and coauthors (2009). How sensitive are the results to the horizontal resolution of the NWF model that is being used to produce the predictor fields? Is it necessary to use a mesoscale model at all? Are precipitation forecasts by statistical downscaling models comparable in quality from the ones by a high-resolution mesoscale model? Which is the influence of the selected predictor combinations in the performance of the statistical downscaling model? Do optimal predictors exist based on phenomenological characteristics of the atmosphere such as convective instability or baroclinicity?

Figure 3 shows the main answers to those questions. This Figure shows the relative operating characteristics curves (ROC) corresponding to precipitation forecasts produced by several models. The models that are closest to the (0,1) point in this curve are the ones that yield better forecasts, according to this skill score. The models are better than a random forecast when they depart from the diagonal line. In this Figure, the X-axis represents the False Alarm Rate or Probability of False Detection (POFD) and the Y-axis represents the Probability of Detection (POD), also called Hit Rate. They are computed after building a verification or contingency table for several precipitation thresholds (0, 0.02, 0.05, 0.08, 0.1, 0.2, 0.3, 0.5, 1 and 1.5 mm h^{-1}). POFD measures the fraction of false alarms given the event did not occur, but was forecast. POD or hit rate measures the fraction of observed events that were correctly forecast.

Figure 3 shows (left panel) the results corresponding to one of the simplest models, H7-3. This model uses as predictors in the analogues+CCA procedure the geopotential heights at two different isobaric levels (700 and 300 hPa). This H7-3 model is the one with the poorest performance, despite the fact that it should theoretically capture the baroclinic structure of the atmosphere. One of the ways to improve this model could theoretically be to complement the information given to the model by means of some indices related to convective instability. Therefore, the left panel of Figure 3 shows the results obtained by two additional models, KI-VWS-H7-3 and SI-VWS-H7-3. These models also used the analogue search followed by CCA in the subset of the best analogues found. These models use as predictors the K and the Showalter instability indices, the vertical wind shear between 700 and 300 hPa and the geopotential height fields at 700 and 300 hPa, with the idea that they will be able to forecast precipitation systems of baroclinic and convective origin. Results show that the use of convective instability indices help a little bit in terms of precipitation forecast, particularly at the low values of POFD (middle and high intensity events). However, the ROC curves corresponding to these models do not clearly separate from the one corresponding to H3-7 alone. It has to be considered that those models were selected on the basis of phenomenological evaluations such as "precipitation is either convective or baroclinic". Those considerations might or might not be strictly true. In the same left panel of Figure 3, the line corresponding to D3-MM5 shows the ROC curve corresponding to precipitation given by the same kind of statistical downscaling model (AnCCA) using the information provided by the mesoscale model at the highest horizontal resolution used (9 km x 9km) in this chapter. This model performs better than previous statistical downscaling models. The statistical downscaling model is able to produce better forecasts when it is allowed to search the output from all the available information without initial constraints.

During early phases of the study it was considered that the moisture balance equation could yield a sound theoretical basis for a forecasting scheme. That is, the moisture conservation equation at the model's grid point is

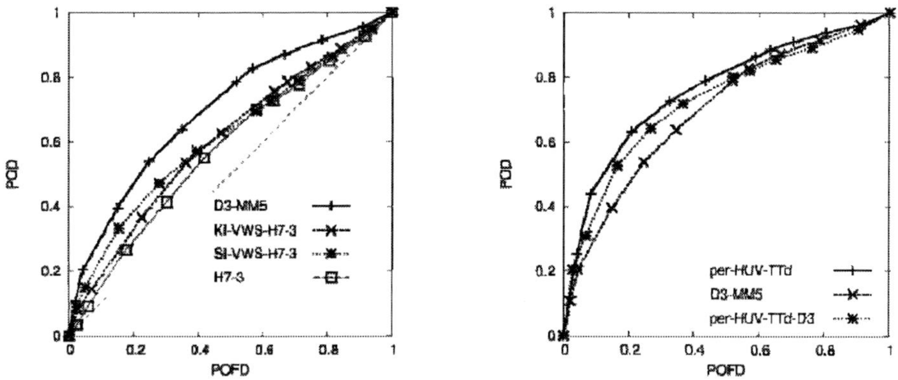

Figure 3. Relative operating characteristics curves corresponding to different methods used to estimate 6h precipitation. Each one of the ROC curves represents the average of POFD and POD at the six observatories (Alg, Lam, Gal, Son, Sif, Abu). On the left panel, D3-MM5 represents the precipitation forecast by means of a AnCCA statistical downscaling model using as predictors geopotential height, temperature and zonal and meridional wind at 1000, 850, 700, 500 and 300 hPa from the integration with MM5 at the highest resolution used in this chapter (9km) over the area. KI-VWS-H7-3 represents a model based on analogues with the K instability index, the vertical wind shear and vertical variation of geopotential between 700 hPa and 300 hPa) as predictors. SI-VWS-H7-3 represents a similar model, but using the Showalter instability index instead of the K index. H7-3 represents a model where the vertical structure of the geopotential height field is considered alone (geopotential height at 700 and 300 hPa are the only predictors). On the right panel, the models per-HUV-TTd and per-HUV-TTd-D3 represent the best performing models, using as predictors geopotential and components of the wind at several isobaric levels, the structure of the moisture field over the sites (given by temperature and dew-point temperature) as obtained from the ECMWF analysis (per-HUV-TTd) or the MM5 integration (per-HUV-TTd-D3) plus persistence, given by the precipitation observations 6 hours before. In this panel the line D3-MM5 is the same as in the previous case and is left for reference.

$$\frac{\partial W_c}{\partial t} + \vec{\nabla} \bullet \vec{Q}_c + \frac{\partial W}{\partial t} + \vec{\nabla} \bullet \vec{Q} = E - P, \qquad (2)$$

with W and W_c the vertically integrated water vapour and condensed water (as water or ice), $\vec{Q}_c$ and $\vec{Q}$ the vertically integrated transports of condensed water and water vapour, E evaporation at the surface and P precipitation. The MM5 model run produces most of the fields in this equation. Therefore, the equation could be used as the basis of a statistical linear downscaling model for the prediction of precipitation (P^*) as follows, assuming that the error η representing the lack of moisture balance by the model will be small.

$$P^* = P + \eta = E - \frac{\partial W_c}{\partial t} - \vec{\nabla} \bullet \vec{Q}_c - \frac{\partial W}{\partial t} - \vec{\nabla} \bullet \vec{Q} + \eta \tag{3}$$

However, evaluation of the linear statistical models based on this consideration showed that the MM5 model did not properly respect the balance equation and statistical linear downscaling models were not better than the previously showed ones based on geopotential or convective instability.

The next step of the study by Fernández-Ferrero et al. (2009) consisted in the search of a set of predictors that allowed a better performance of the downscaling model. It was found that the statistical downscaling model using the analogues + CCA (AnCCA) strategy that performed the best of all the checked ones was a model called per-HUV-TTd. In this model, part of the rainfall forecast derives from previous rainfall observations (this corresponds to the persistence part of the model). Additionally, the HUV term represents the contribution from the atmospheric circulation (geopotential height, H and zonal and meridional wind components, U and V) at several vertical isobaric levels (1000, 850, 700, 500 and 300 hPa) from atmospheric analysis. The TTd block provides moisture information, that is, temperature and dew point temperature at 850, 700 and 500 hPa at the grid point closest to the place were the forecast is being issued. We tested two models, per-HUV-TTd, which uses as predictors data from ERA40 and per-HUV-TTd-D3, which uses the predictors resulting from the integration with the mesoscale model at the highest resolution but without assimilation of observations. Both statistical downscaling models yield better forecasts than the previously identified D3-MM5 model, so that information on moisture and persistence help in the precipitation forecast. Additionally, both models yield better forecasts than the precipitation predicted by the mesoscale model at the highest resolution used. This result also holds in terms of correlation and variance fraction too. The model using as predictors data from ERA40 works better than the one using as predictors fields from the MM5 integration. This reflects the importance of data assimilation used in ERA40 but not in the MM5 run for precise analysis of atmospheric moisture. Still, the fact that the output from the statistical downscaling model per-HUV-TTd-D3 is better than the raw precipitation given by the model output indicates that downscaling yields an added value to the output of the high-resolution model.

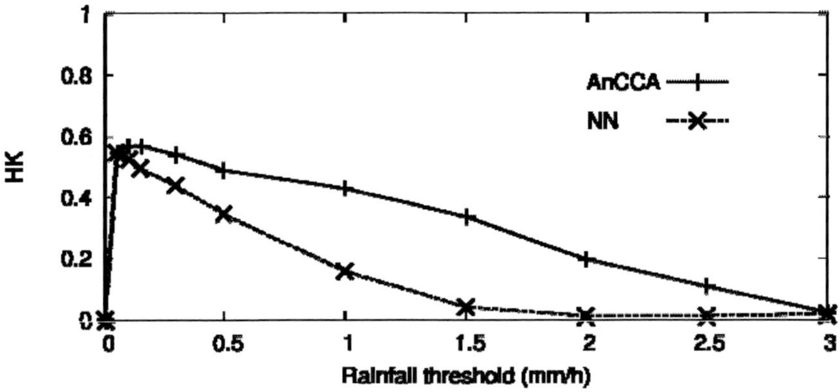

Figure 4. Hanssen-Kuipers skill score corresponding to the best AnCCA or neural networks based models using the same set of predictors for different rainfall thresholds.

After the best combination of predictors was obtained, we wanted to compare the performance of our AnCCA downscaling model with the best one based on neural networks that we could develop. Figure 4 shows the performance of the AnCCA and the best NN model using the same set of predictors for different rainfall thresholds.

It can be seen that, except for the very low precipitation thresholds, AnCCA model works better than the best NN model for intense precipitation events. The values of the Hanssen-Kuipers skill score from the AnCCA and NN methods are statistically different with a 95% confidence level above the 0.5 mm h^{-1} rainfall threshold, according to a bootstrap test leading to estimations of their error bars.

Probabilistic QPF at Short Range (6 Hours)

The best of the statistical downscaling models identified in the previous study (Fernández-Ferrero et al., 2009), consisting of a search of analogues plus a CCA in the subset of the best analogues with persistence, geopotential height, zonal and meridional components of the wind and temperature and dew point temperature over the point in which the forecast is being issued (per-HUV-TTd as shown in Figure 3) was the basis of the formulation of a probabilistic model for rainfall forecast at a short lead time (6 hours) over the rain gauges in Bilbao (Abu, Alg, Gal, Lam, Sif and Son), as shown in Figure 1. The probabilistic models built ranged from a simple frequentist one, based on the probability of exceedance given by a subset of analogues (A) to three Bayesian models encompassing different levels of sophistication (Fernández-Ferrero et al., 2010). The first Bayesian model (B) uses as prior the output from the A model and as likelihood an experimental fit of a normal (lognormal for low precipitation thresholds) distribution, with parameters equal to the observed ones. The second Bayesian model (BE) involves including the observed precipitation in the computation of the distance used to derive the likelihood, and the analytical form of the likelihood corresponds to a normal distribution. Finally, the BI model is similar to the BE model, but it uses a fit to a potential law for the likelihood.

Table 2 shows the Brier score and the Brier skill score (BSS), computed with respect to the climatological frequency corresponding to each of the precipitation thresholds used in the study, according to forecasts by each of the probabilistic models (A, B, BE and BI). The Brier score in Table 2 shows that the relative performance of all the models is similar for all the thresholds (the values are very similar at each row). The data corresponding to the BSS shows that all the models provide forecasts better than the climatology for rainfall thresholds up to 1.5 mm h^{-1}. All the models are unable to produce better forecasts than the climatology (observed probability of the event) for the higher thresholds. In order to understand the behaviour of the individual models, the Brier score was partitioned in reliability and resolution components. In terms of reliability, the B model is the worst one, whilst BI is the best one. According to resolution, the Bayesian models (B, BE and BI) show lower (worse) values of the resolution than the frequentist model A.

Figure 5 shows, stratified by the observed precipitation threshold, the Ranked Probability Score (RPS) obtained from all the probabilistic precipitation forecasts. The box and whiskers (minimum, 25th quantile, median, 75th quantile and maximum) indicate the distribution of sampled RPS values for all the precipitation thresholds. The analysis of the distribution of RPS values shows that A yields better forecasts most of the time. However, when the forecast of the events is difficult, the Bayesian models show better RPS values than the A model. For lowest precipitation thresholds, the A model works slightly better, but, overall, all the models work properly and the spread of the RPS values is very low on those low thresholds (up to 0.5 mm h^{-1}). From 0.5 mm h^{-1} onwards, the frequentist model systematically shows a worst forecast ability (as measured, for instance, by the median of the RPS values). In general, Bayesian models show a higher value of the median of the RPS, with BE and BI showing a smaller spread of the RPS for "difficult" precipitation thresholds such as 2 mm h^{-1}, that are still better predicted by these models BE and BI than by climatology, according to the BSS values shown in Table 2. The deterioration of the performance of the models for high precipitation thresholds can be either due to the inherent difficulty of the forecast or to the short record used in the study, and further research is needed to elucidate this aspect.

Table 2. Brier score (BS) and Brier skill score (BSS) for each of the models and thresholds used in the probabilistic precipitation forecast

Threshold	BS				BSS			
mm/h	A	B	BE	BI	A	B	BE	BI
0.05	0.10	0.11	0.10	0.10	0.47	0.39	0.45	0.46
0.10	0.09	0.10	0.10	0.09	0.44	0.37	0.41	0.43
0.25	0.07	0.08	0.08	0.07	0.38	0.32	0.33	0.36
0.50	0.05	0.06	0.05	0.05	0.30	0.24	0.26	0.28
0.75	0.04	0.04	0.04	0.04	0.24	0.18	0.21	0.22
1.00	0.03	0.03	0.03	0.03	0.18	0.13	0.15	0.16
1.25	0.02	0.02	0.02	0.02	0.14	0.09	0.11	0.11
1.50	0.02	0.02	0.02	0.02	0.11	0.05	0.09	0.09
2.00	0.01	0.01	0.01	0.01	0.03	-0.04	0.01	0.01
3.00	0.00	0.00	0.00	0.00	-0.13	-0.24	-0.14	-0.15
4.00	0.00	0.00	0.00	0.00	-0.45	-0.59	-0.45	-0.48

Downscaling of Daily Precipitation at the Ebro Valley

In the next example of application of statistical downscaling for precipitation forecast, taken from Ibarra-Berastegi et al. (2011), the statistical downscaling model proceeds through a first step (preselection of a subset of 300 analogues) that is followed by a stage of multiple linear regression (MLR) or Random Forests (RF). For comparison purposes, the output from a subset of analogues (Analogues) is also shown. The target variable is daily precipitation at two sites in the middle (Zaragoza, Z) and the end of the Ebro Valley (Tortosa, TO) as recorded in the ECA repository. Figure 6 shows the Taylor diagrams used to compare the performance of all the methods.

In general, at the daily time scale, all the methods underestimate the standard deviation of the precipitation. The case of MLR is an exception that deserves a special commentary. In all cases the downscaling model has been trained using data from the early record in ERA40 reanalysis. Then, the statistical downscaling model has also been used with anomalies derived from ERA Interim. In the case of MLR, the bias between both models produces an abrupt change in the standard deviation of downscaled precipitation. Conversely, the use of RF makes the method less sensitive to a bias between the training and the production dataset.

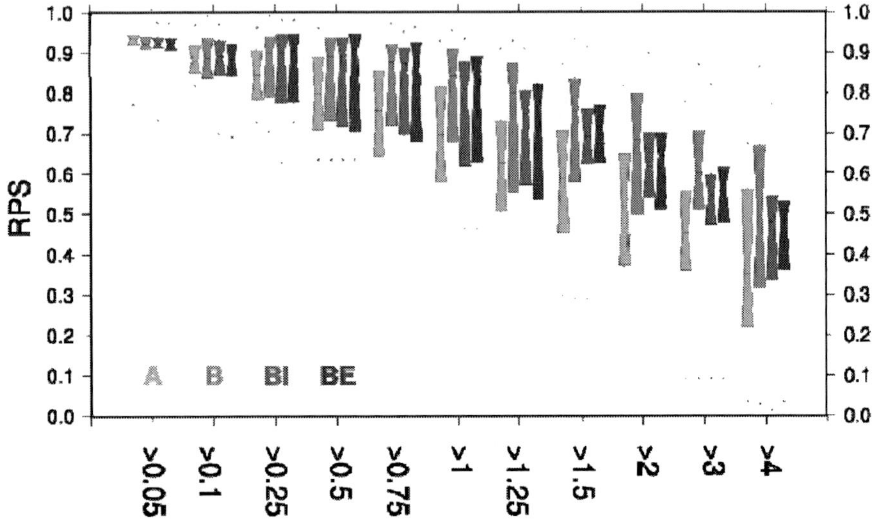

Figure 5. Ranked probability scores corresponding to the probabilistic forecasts produced by the four probabilistic models used, shown as box and whiskers. A perfect probabilistic forecast would correspond to a RPS value of 1 as used in this chapter. From left to right, the thresholds correspond to the values 0.05, 0.1, 0.25, 0.50, 0.75, 1, 1.25, 1.50, 2, 3 and 4 mm h^{-1}.

In terms of correlation, the values are relatively low, but given the size of the test data set (1826 cases spanning the period 1997-2001), they are significantly different from zero with a confidence level over 99.9%, according to a Monte Carlo test considering the autocorrelation of the observed series. A bootstrap test on the difference of the performance of the precipitation forecast resulting from all methods shows that the correlation coefficient or root mean square error skill scores of all the models (Analogues, MLR-ERA40 and RF-ERA40) do not differ with a confidence of 95%. It can therefore be concluded that for precipitation forecasts, the use of random forests plus analogues or multiple linear regression plus analogues does not improve the performance of the analogues stage of the downscaling method. The analysis of the relative importance of the variables in the downscaling process shows (Ibarra-Berastegi et al., 2011) that the results are sensitive to the first EOF of relative humidity, the first EOF of meridional wind speed and the second EOF of dew point temperature at the surface (inside the planetary boundary layer).

CONCLUSION

An objective analysis of the clustering of synoptic situations and lightning by means of SOM analysis shows that there exist three meteorological patterns that affect rainfall yield during lightning vents and these results are in agreement with previous subjective classifications (Ezcurra et al., 2002). These patterns show a remarkable seasonality and the use of the automatic classification procedure allows finding a proper physical interpretation of the association of lightning, meteorological situation and rainfall. The first cluster (node in the SOM matrix) is formed by events that can be interpreted in terms of baroclinic disturbances affecting the area, with a substantial convergence of moisture transports from the ocean and high rainfall yield rate. The second cluster is formed by events showing a characteristic negative thermal anomaly at high levels in the troposphere, with substantial convective instability covering wide areas. Finally, the last node of the SOM is linked to events that show weaker geopotential height anomalies, intense convective instability covering limited areas and weaker convergence of moisture transports, with reduced precipitation. Despite the agreement of classification results with the physical interpretation, the quantitative relationship between lightning and precipitation is still an open problem.

Results from the comparison of precipitation forecasts with a lead time of six hours by a downscaling model and a mesoscale model that does not assimilate observations have shown that a properly designed statistical downscaling model produces better rainfall forecasts. We are currently performing a long integration by means of the WRF/ARW model performing 3DVAR assimilation of observations every six hours. The role played by 3DVAR data assimilation will thus be checked in future works.

The comparison of the results produced by the analogue + CCA (AnCCA) and the neural network based methods shows that in general the AnCCA method produces better forecasts for medium intensity rainfall quantities, whilst they produce very similar results for low precipitation events.

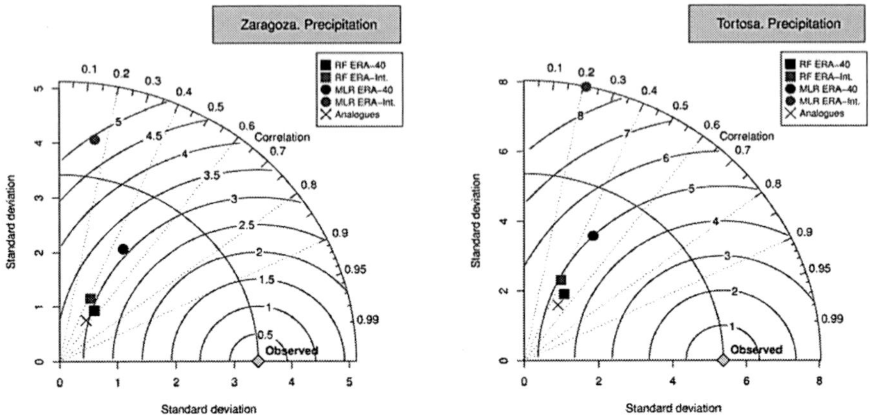

Figure 6. Taylor diagrams corresponding to the daily precipitation values in Zaragoza (left) and Tortosa (right), as evaluated during the period 1989-2001. For each kind of combined model (analogues plus random forest or analogues plus multiple linear regression), the results corresponding to predictor fields from ERA40 and ERA Interim are compared together with the results corresponding to a simple analogues model.

The statistical downscaling model is able to produce better precipitation forecasts if it is given all the information regarding predictors than if there exist previous informed guesses (baroclinic activity, convective instability and so on) that are used to previously condition the information received by the statistical downscaling algorithm. Ideally optimal methods (such as the use of the equations that represent the hydrological cycle) do not necessarily perform better, due to the lack of balance of the tested numerical model. This conclusion might change with new models, such as with the use of WRF/ARW, but this is still under development.

The use in the statistical downscaling model of the previously observed precipitation significantly enhances the quality of the precipitation forecasts at a short term. It would be interesting to check in future works ways to involve additional sources of information such as radar in order to combine all the predictors, so that the statistical model was able to use all the sources of information in order to enhance its quantitative skill scores.

The extension of the AnCCA (or analogues-based models) to probabilistic forecast is quite simple and results in this chapter show that these models are able to produce quantitative probabilistic forecasts that are more skilful than climatology for a range of precipitation thresholds, except the ones corresponding to most intense events. This is a critical problem, since these events are the most important from the point of view of an operational application of these methods. In general, this is due to the well-known fact that analogue-based methods are not able to produce forecasts that have not already been observed (Obled et al., 2002; Imbert and Benestad, 2005; Benestad 2010). There are currently under study ways to circumvent this problem. One of the possibilities that are currently been used is the recalibration of the forecasts by mean of analogues (Benestad, 2010). A second option is to enlarge the database of observations by means of reforecast datasets in order to produce longer time series that better sample the historical record and allow a better performance of numerical weather forecast model correction schemes such as MOS or analogues (Hamill et al., 2004; Hamill and Whitaker, 2006).

Despite the simplicity of the analogue based approach to downscaling, it has been shown that it is very competitive in front of alternative techniques such as neural networks for short term or Random Forests. This result has already been identified in previous studies (Zorita and von Storch, 1999). However, it is to be seen whether the technique can be used free of error in the downscaling of climate scenarios, since some studies (Frías et al., 2006) have shown that the ability of the method to simulate a changing climate by means of analogues depends to some extent on the study area.

References

Altava-Ortiz, V., Barrera, A., Llasat, C., Prat, M. A., Gibergans-Báguena, J. and Barnolas, M. (2006) "Application of the MM5 and the analogous method to heavy rainfall event, the case of 16–18 October 2003 in Catalonia (NE Spain)." *Advances in Geosciences*, Vol. 7, pp. 313-319.

Bauman, W. H., Kaplan, M.L., Businger, S. (1997) "Nowcasting convective activity for space shuttle landings during easterly flow regimes." *Weather and Forecasting*, Vol. 12, No. 1, pp. 78–107.

Benestad, R. E., Hanssen-Bauer, I., and Chen, D. (2008): "Empirical-Statistical Downscaling", World Scientific Publishers, Singapore.

Benestad, R.E. (2010) "Downscaling precipitation extremes: correction of analog models through PDF predictions", *Theoretical and Applied Climatology*, Vol. 100, No. 1-2, pp. 1-21.

Bishop, C.M., (1995) "Neural Networks for Pattern Recognition." Oxford University Press, Oxford, England.

Bohren, C.F., and Albrecht, B.A., (1998) "Atmospheric Thermodynamics", Oxford University Press, Oxford.

Boudevillain, B., Andrieu, H. and Chaumerliac, N. (2006) "Evaluation of RadVil, a radar-based very short-term rainfall forecasting model." *Journal of Hydrometeorology*, Vol. 7, No. 1, pp. 178–189.

Bowden, G.J., Maier, H.R. and Dandy, G.C., (2001) "Optimal division of data for neural network models in water resources applications." *Water Resources Research*, Vol. 38, No. 2, pp. 1010.

Breiman, L. (2001) "Random forests", *Machine Learning*, Vol. 45, No. 1, pp. 5–32.

Bretherton, C. S., Smith, C., and Wallace, J.M., (1992) "An Intercomparison of Methods for Finding Coupled Patterns in Climate Data." *Journal of Climate*, Vol. 5, No. 6, pp. 541–560.

Busuioc, A., and von Storch, H., (2003) "Conditional stochastic model for generating daily precipitation time series." *Climate Research*, Vol. 24, No. 2, pp. 181–195.

Cavazos, T., (1999) "Large-scale circulation anomalies conductive to extreme precipitation events and derivation of daily rainfall in north-eastern in Mexico and south-eastern Texas." *Journal of Climate* Vol. 12, No. 5, pp. 1506–1523.

Cavazos, T., (2000) "Using self-organizing maps to investigate extreme climate event: an application to wintertime in the Balkans." *Journal of Climate, Vol.* 13, No. 10, pp. 1718–1732.

Charney, J. G., Fjortoft, J., and von Neumann, J. (1950) "Numerical integration of the barotropic vorticity equation." *Tellus*, Vol . 2, No. 4, pp. 237-254.

Diomede, T., Nerozzi, F., Paccagnella, T., and Todini, E. (2008) "The use of meteorological analogues to account for LAM QPF uncertainty." *Hydrology and Earth System Sciences*, Vol. 12, No. 1, pp. 141–157.

Djurić, D., (1994) "Weather Analysis", Prentice-Hall, New Jersey.

Ebert, E. E., Wilson, L. J., Brown, B. G., Nurmi, P., Brooks, H. E., Bally, J. and Jaeneke, M. (2004) "Verification of nowcasts from the WWRP Sydney 2000 forecast demonstration." *Weather and Forecasting*, Vol. 19, No. 1, pp. 73-96.

Ezcurra, A., Areitio, J., and Herrero, I. (2002) "Relationship between Cloud-to-Ground Lightning and Surface Rainfall During 1992-1996 in the Spanish Basque Country". *Atmospheric Research*, Vol. 61, No. 3, pp. 239-250.

Ezcurra, A., Sáenz, J., Ibarra-Berastegi, G. and Areitio, J. (2008) "Rainfall yield characteristics of electrical storm observed in the Spanish Basque

Country during the period 1992-1996." Atmospheric Research, Vol. 89, No. 3, pp. 233-242.

Fernández, J. and Sáenz, J., (2003) "Improved field reconstruction with the analog method: searching the CCA space." *Climate Research*, Vol. 24, pp. 199-213.

Fernández-Ferrero, A., Sáenz, J., Ibarra-Berastegi, G. and Fernández, J. (2009) "Evaluation of statistical downscaling in short range precipitation forecasting." *Atmospheric Research,* Vol. 94, No. 3, pp. 448-461.

Fernández-Ferrero, A., Sáenz, J. and Ibarra-Berastegi, G. (2010) "Comparison of the performance of different analog-based Bayesian probabilistic precipitation forecasts over Bilbao, Spain." *Monthly Weather Review,* Vol. 138, No. 8, pp. 3107-3119.

Font-Tullot, I. (2000) "Climatología de España y Portugal." Universidad de Salamanca, Salamanca, Spain.

Fox, N. I. and Wikle, C.K., (2005) "A Bayesian Quantitative Precipitation Nowcast scheme." *Weather and Forecasting*, Vol. 20, No. 3, pp. 264–275.

Frías, D, Zorita, E., Fernández, J., and Rodríguez-Puebla, C., (2006) "Testing statistical downscaling methods in simulated climates." *Geophysical Research Letters*, Vol. 33, L19807.

Golding, B. W. (1998) "Nimrod: a system for generating automated very short range forecasts." *Meteorological Applications*, Vol. 5, No. 1, pp. 1–16.

Grell, G. A., Dudhia, J., Stauffer, D. R., (1994) "A description of the fifth generation Penn State/NCAR mesoscale model (MM5)." Tech. Rep. NCAR/TN-398+STR, National Center for Atmospheric Research, Boulder, CO, USA.

Grömping, U., (2009) "Variable Importance Assessment in Regression: Linear Regression versus Random Forest." *The American Statiscian,* Vol. 63, pp. 308-319.

Gutiérrez, J. M., Cofiño, A. S., Cano, R., and Rodríguez, M. A. (2004) "Clustering methods in statistical downscaling for short-range weather forecasts." *Monthly Weather Review*, Vol. 132, No. 9, pp. 2169–2183.

Gutiérrez, J.M., Cano, R., Cofiño, A.S., and Sordo, C., (2005) "Analisys and downscaling multi-model seasonal forecast in Peru using self-organizing maps." *Tellus,* Vol. 57A, No. 3, pp. 435–447.

Hamill, T. M., Whitaker, J. S., and Wei, X., (2004) "Ensemble reforecasting: improving medium-range forecast skill using retrospective forecasts." *Monthly Weather Review*, Vol. 132, No. 6, pp. 1434–1447.

Hamill, T. M., and Whitaker, J. S., (2006) "Probabilistic quantitative precipitation forecasts based on reforecast analogs: Theory and application." *Monthly Weather Review*, Vol. 134, No. 11, pp. 3209–3229.

Hart, K. A., Steenburgh, W. J., Onton, D. J. and Siffert, A. J. (2004) "An evaluation of mesoscale-model-based model output statistics (MOS) during the 2002 Olympic and Paralympic Winter Games." *Weather and Forecasting*, Vol. 19, No. 2, pp. 200-218.

Hastie, T., Tibshirani, R., and Friedman, J., (2002) "The elements of statistical learning. Data mining, inference and prediction." Springer-Verlag, Berlin.

Haykin, S., (1994) "Neural Networks: A Comprehensive Foundation." Macmillan Publishing, New York.

Hewitson, B.C., and Crane, R.G., (2002) "Self-organizing maps: applications to synoptic climatology." *Climate Research* Vol. 22, No. 1, pp. 13–26.

Ibarra-Berastegi, G., Elías, A., Barona, A., Sáenz, J., Ezcurra, A. and Díaz de Argandoñia, J., (2008) "From diagnosis to prognosis for forecasting air pollution using neural networks: air pollution monitoring in Bilbao." *Environmental Modelling and Software*, Vol. 23, pp. 622–637.

Ibarra-Berastegi, G., Saénz, J., Ezcurra, A., Elías, A., Diaz Argandoña, J., and Errasti, I., (2011) "Downscaling of surface moisture flux and precipitation in the Ebro Valley (Spain) using analogues and analogues followed by random forests and multiple linear regression", *Hydrology and Earth System Sciences*, Vol. 15, No. 6, pp. 1895-1907.

Imbert, A., and Benestad, R.E. (2005) "An improvement of analog model strategy for more reliable local climate change scenarios." *Theoretical and Applied Climatology*, Vol. 82, No. 3-4, pp. 245-255.

Kidson, J. W. and Thompson, C. S. (1998) "A comparison of statistical and model-based downscaling techniques for estimating local climate variations." *Journal of Climate*, Vol. 11, No. 4, pp. 735–753.

Kirstetter, P. E., Delrieu G., Boudevillain B. and Obled C. (2010) "Toward an error model for radar quantitative precipitation estimation in the Cévennes–Vivarais region, France." *Journal of Hydrology*, Vol. 394, No. 1-2, pp. 28-41.

Klein Tank, A.M.G., Wijngaard, J. B., Können, G. P., Böhm, R., Demarée, G., Gocheva, A., Mileta, M., Pashiardis, S., Hejkrlik, L., Kern-Hansen, C., Heino, R., Bessemoulin, P., Müller-Westermeier, G., Tzanakou, M., Szalai, S., Pálsdóttir, T., Fitzgerald, D., Rubin, S., Capaldo, M., Maugeri, M., Leitass, A., Bukantis, A., Aberfeld, R., van Engelen, A. F. V., Forland, E., Mietus, M., Coelho, F., Mares, C., Razuvaev, V., Nieplova, E., Cegnar, T., Antonio López, J., Dahlström, B., Moberg, A., Kirchhofer,

W., Ceylan, A., Pachaliuk, O., Alexander, L. V. and Petrovic, P. (2002), "Daily dataset of 20th-century surface air temperature and precipitation series for the European Climate Assessment." *International Journal of Climatology*, Vol. 22, No. 12, pp. 1441–1453.

Kohonen, T., (1982) "Self-organized formation of topologically correct feature maps." *Biological Cybernetics*, Vol. 43, pp. 59–69.

Kotroni, V., and Lagouvardos, K. (2004) "Evaluation of MM5 high-resolution real-time forecasts over the urban area of Athens, Greece." *Journal of Applied Meteorology*, Vol. 43, No. 11, pp. 1666–1678.

Kruizinga, S., and Murphy, A.H., (1983) "Use of an analogue procedure to formulate objective probabilistic temperature forecasts in The Netherlands." *Monthly Weather Review*, Vol. 111, No. 11, pp. 2244–2254.

Krzysztofowicz, R., (1998) "Probabilistic hydrometeorological forecasts: Toward a new era in operational forecasting." *Bulletin of the American Meteorological Society*, Vol. 79, No. 2, pp. 243–251.

Kuligowski, R. J., and Barros, A. P., (1998) "Localized precipitation forecasts from a numerical weather prediction model using artificial neural networks." *Weather and Forecasting*, Vol. 13, No. 4, pp. 1194–1204.

Lack, S. A. and Fox, N.I. (2007) "An examination of the effect of wind-drift on radar-derived surface rainfall estimations." *Atmospheric Research*, Vol. 85, No. 2, pp. 217-229.

Liaw, A., and Wiener, M. (2002) "Classification and Regression by randomForest." *R News*, Vol. 2, No. 3, pp. 18-22.

Lorenz, E. N., (1969) "Atmospheric Predictability as Revealed by Naturally Occurring Analogues." *Journal of Atmospheric Sciences*, Vol. 26, No. 4, pp. 636–646.

Lu, W.Z., Fan, H.Y., Leung, A.Y.T., and Wong, J.C.K., (2002) "Analysis of pollutant levels in central Hong Kong applying neural network method with particle swarm optimization." *Environmental Monitoring and Assessment*, Vol. 79, No. 3, pp. 217–230.

Maier, H.R. and Dandy, G.C., (2000) "Neural networks for the prediction and forecasting of water resources variables: a review of modelling issues and applications." *Environmental Modelling and Software*, Vol. 15, No. 1, pp. 101–124.

Maier, H.R., Morgan, N., and Chow, C.W.K., (2004) "Use of artificial neural networks for predicting optimal alum doses and treated water quality parameters." *Environmental Modelling and Software*, Vol. 19, No. 5, pp. 485–494.

Marty, R., Zin, I., and Obled, C., (2008) "On adapting PQPFs to fit hydrological needs: The case of flash flood forecasting." *Atmospheric Science Letters*, Vol. 9, No. 2, pp. 73–79.

Marzban, C., and Stumpf, G. J. (1998) "A neural network for damaging wind prediction." *Weather and Forecasting*, Vol. 13, No. 1, pp. 151-163.

Masters, T., (1993) "Practical Neural Network Recipes in C++.", Academic Press, San Diego, USA.

Matulla, C., Zhang, X., Wang, X. L., Wang, J., Zorita, E., Wagner, S., von Storch, H. (2008) "Influence of similarity measures on the performance of the analog method for downscaling daily precipitation." *Climate Dynamics*, Vol. 30, No. 2-3, pp. 133-144.

McCann, D. W., (1992) "A neural network short-term forecast of significant thunderstorms." *Weather and Forecasting*, Vol. 7, No. 3, pp. 525–534.

Nicolis, C., (1998) "Atmospheric Analogs and Recurrence Time Statistics: Toward a Dynamical Formulation." *Journal of the Atmospheric Sciences*, Vol. 55, No. 3, pp. 465–475.

Obled, C., Bontron, G., and Garcon, R. (2002) "Quantitative precipitation forecasts: a statistical adaptation of model outputs through an analogues sorting approach." *Atmospheric Research*, Vol. 63, No. 3-4, pp. 303-324,

Patterson, D., (1996) "Artificial Neural Networks." Prentice Hall, Singapore.

Roebber, P. J., Bosart, L. F. (1998) "The sensitivity of precipitation to circulation details. Part I: An analysis of regional analogs." *Monthly Weather Review*, Vol. 126, No. 2, pp. 437–455.

Shao, J., (1998) "Improving nowcasts of road surface temperature by a backpropagation neural network." *Weather and Forecasting*, Vol. 13, No. 1, pp. 164–171.

Sheridan, S. C. and Cameron, C. L. (2011) "The self-organizing map in synoptic climatological research." *Progress in Physical Geography,* Vol. 35, No. 1, pp. 109-119.

Siroky, D.S. (2009) "Navigating Random Forests and related advances in algorithmic modeling." *Statistics Surveys*, Vol. 3, pp. 147-163.

Smith, K. T. and Austin, G. L. (2000) "Nowcasting precipitation - a proposal for a way forward." *Journal of Hydrology*, Vol. 239, No. 1–4, pp. 34–45.

Stephenson, D. B., Coelho, C. A. S., Doblas-Reyes, F. J., and Balmaseda, M., (2005) "Forecast assimilation: A unified framework for the combination of multi-model weather and climate predictions." *Tellus*, Vol. 57A, No. 3, pp. 253–264.

Thielen, J., Boudevillain, B. and Andrieu, H. (2000) "A radar data based short-term rainfall prediction model for urban areas—A simulation using mesoscale meteorological modelling." *Journal of Hydrology*, Vol. 239, No. 1-4, pp. 97–114.

Toth, Z., (1991) "Intercomparison of circulation similarity measures." *Monthly Weather Review*, Vol. 119, pp. 55–64.

Trenberth, K.E., and Guillemot, C.J., (1995) "Evaluation of the Global Atmospheric Moisture Budget As Seen From Analyses." *Journal of Climate*, Vol. 8, No. 9, pp. 2255—2272.

Uppala, S.M., Kallberg, P.W., Simmons, A.J., Andrae, U., da Costa Bechtold, V., Fiorino, M., Gibson, J.K., Haseler, J., Hernández, A., Kelly, G.A., Li, X., Onogi, K., Saarinen, S., Sokka, N., Allan, R.P., Andersson, E., Arpe, K., Balmaseda, M.A., Beljaars, A.C.M., van de Berg, L., Bidlot, J., Bormann, N., Caires, S., Chevallier, F., Dethof, A., Dragosavac, M., Fisher, M., Fuentes, M., Hagemann, S., Holm, E., Hoskins, B.J., Isaksen, L., Janssen, P.A.E.M., Jenne, R., McNally, A.P., Mahfouf, J-F., Morcrette, J-J., Rayner, N.A., Saunders, R.W., Simon, P., Sterl, A., Trenberth, K.E., Untch, A., Vasiljevic, D., Viterbo, P., and Woollen, J., (2005) "The ERA-40 Re-analysis." *Quarterly Journal of the Royal Meteorological Society*, Vol. 131, No. 612, pp. 2961-3012.

van den Dool, H. M. (1994) "Searching for analogues, how long must we wait?" Tellus A, Vol. 46, No. 3, pp. 314–324.

Vasiloff, S. V., Howard, K. W., Rabin, R. M., Brooks , H. E., Seo, D. J., Zhang, J., Kitzmiller, D. H., Mullusky, M. G., Krajewski, W. F. , Brandes, E. A., Brown , B. G., Berkowitz, D. S., McGinley, J. A., and Kuligowski, R. J. (2007) "Improving QPE and very short term QPF: an initiative for a community-wide integrated approach." *Bulletin of the American Meteorological Society*, Vol. 88, No. 12, pp. 1899–1911.

Weichert, A. and Bürger, G. (1998) "Linear versus nonlinear techniques in downscaling." *Climate Research*, Vol. 10, No. 2, pp. 83-93.

Wetterhall, F., Halldin, S. and Xu, C. (2005) "Statistical precipitation downscaling in central Sweden with the analogue method." *Journal of Hydrology*, Vol. 306, No. 1-4, pp. 174–190.

Wilby, R.L. and Wigley, T.M.L. (1997) "Downscaling general circulation model output: a review of methods and limitations." *Progress in Physical Geography, Vol.* 21, No. 4, pp. 530-548.

Wilby, R. L., Charles, S. P., Zorita, E., Timbal, B., Whetton, P. and Mearns, L. O. (2004) "Guidelines for use of climate scenarios developed from statistical downscaling methods.", IPCC, Geneva, Switzerland.

Witten, I.H., and Frank, E. (2005) "Data mining. Practical machine learning tools and techniques", 2nd Ed., Elsevier, Oxford, England.

Xiao, Q. and Sun, J., (2007) "Multiple-radar data assimilation and short-range quantitative precipitation forecasting of a squall line observed during IHOP_2002." *Monthly Weather Review*, Vol. 135, Vol. 10, pp. 3381–3404.

Xu, J., Xiao, Q., Gao, X., and S. Sorooshian (2006) "Influence of assimilating rainfall derived from WSR-88D radar on the rainstorm forecasts over the southwestern United States." *Journal of Geophysical Research*, Vol. 111, D13105.

Yates, E., Anquetin, S., Ducrocq, V., Creutin, J. D., Ricard, D. and Chancibault, K. (2006) "Point and areal validation of forecast precipitation fields." *Meteorological Applications*, Vol. 13, No. 1, pp. 1-20.

Yuval, and Hsieh W. W., (2003) "An adaptive nonlinear MOS scheme for precipitation forecasts using neural networks." Weather and Forecasting, Vol. 18, No. 2, pp. 303–310.

Zorita, E. and von Storch, H. (1999) "The analog method as a simple statistical downscaling technique: comparison with more complicated methods." *Journal of Climate*, Vol. 12, No. 8, pp. 2474–2489.

In: Rainfall Forecasting
Editor: T.S.W. Wong, pp. 39-128
ISBN: 978-1-61942-134-9

Chapter 2

EMPIRICAL APPROACHES IN LONG-TERM RAINFALL FORECASTING

Lorna R. Nayagam, Rajesh J. and H. S. Ram Mohan
Department of Atmospheric Sciences,
Cochin University of Science and Technology,
Cochin, Kerala 682 016, India.

ABSTRACT

Long-term forecasting of rainfall for periods longer than seven days over a region of high spatial and temporal variability is a challenging task. Various empirical or numerical attempts have been made to quantify this varying quantity. The effort taken across the globe in the prediction of rainfall has not attained 100% success. In this conjuncture, this chapter is an endeavour to take account of the various empirical techniques used to forecast the rainfall across the globe. The introduction of this chapter gives a description on the historical perspectives, outline of the basic forecasting techniques and the steps in forecasting. In the next section, a better understanding of the time series, some basic quantitative statistics, physical linkage and formation of indices and some common methods in the selection of predictors are described. Subsequent section deal with the various empirical methods used in rainfall forecasting; each explained with the examples from atmospheric science. The section describes both the quantitative (linear and non-linear) and probabilistic methods used to forecast the rainfall. This consists of simple regression, multiple regression, autoregressive integrated moving average (ARIMA) models, multiple power regression, Principal Component regression, Projection Pursuit Regression, Neural Network, Probabilistic forecast from ensembles and parametric model. The techniques discussed so far are used for giving forecasts at a single geographic location at different times. Other than these, the prediction at a spatial scale is common nowadays. Principal Component Analysis (PCA), Canonical Correlation analysis (CCA) and Numerical Weather Prediction (NWP) are the commonly used techniques for the forecasts at spatial fields. Excepting NWP, the empirical methods used in spatial prediction and their forecast verification methods are also discussed. The various measures used in the evaluation and correction of forecasts, the challenges and problems in issuing long-term forecasts, various biases, limitations and common practices to minimize errors are also discussed.

Keywords: Autocorrelation, Analysis of Variance, ARIMA model, Autocorrelation Function, Canonical vectors, Canonical correlation, Canonical Correlation Analysis, Coefficient of determination, Cross validation, Data transformation, Durbin Watson, Eigen value, Eigen vectors, Ensemble forecasts, Evaluation statistics, Forecast verification, Forward and backward passes, Intercorrelation, Least Square Estimation, Multiple Power regression, Multiple Linear regression, Neural Network, Neurons, Optimization, Parametric model, Partial Autocorrelation Function, Portmanteau tests, Principal Component Analysis, Principal Component regression, Probabilistic forecasts, Projection Pursuit Regression, Quantitative forecasts, Regression coefficients, Simple regression, Sliding correlation, Stationary and Non-stationary series, Training and test period, Variance Inflation Factor.

List of Abbreviations

ABSE	Absolute error
ACC	Anonamly Correlation Coefficient
ACF	Autocorrelation Function
AIC	Akaike's Information Criterion
AIR	All India Rainfall
ANN	Artificial Neural Network
ANOVA	Analysis of Variance
AR	Autoregressive
ARIMA	Autoregressive Integrated Moving Average

ARMA	Autoregressive Moving Average
AUC	Area Under Curve
BoM	Bureau of Meteorology
CCA	Canonical Correlation Analysis
df	Degrees of Freedom
DW	Durbin Watson
EBP	Error-Back-Propagation
EOF	Empirical Orthogonal Function
FAR	False Alarm Rate
FFNN	Feed-Forward Neural Network
HS	Hit score
HSS	Heidke Skill Score
ISMR	Indian Summer Monsoon rainfall
KSMR	Kerala Summer Monsoon Rainfall
L	Likelihood
LR	Linear regression
LRF	Long-Range Forecast
MA	Moving Average
MAE	Mean Absolute error
MLR	Multiple Linear Regression
MS	Mean Square
MSD	Mean Squared Deviation
MSE	Mean Squared Error
MSESS	MSE Skill Score
MSLP	Mean Sea Level Pressure
MSR	Regression Mean Square
MST	Total Mean Square
NAO	North Atlantic Oscillation
NCC	National Climate Centre
NPO	North Pacific Oscillation
NWP	Numerical Weather Prediction
PACF	Partial Autocorrelation Function
PCA	Principal Component Analysis
PCR	Principal Component Regression
PCs	Principal Components
POD	Probability of Detection
PP	Projection Pursuit
PPR	Projection Pursuit Regression
RMSE	Root Mean Square Error

ROC	Relative Operating Characteristic
s.e.	Standard Error
SARIMA	Seasonal ARIMA model
SBIC	Schwarz Bayesian Information Criterion
SO	Southern Oscillation
SS	Sum of Squares
SSE	Sum Squared Error
SSR	Explained Sum of Squares
SST	Sea Surface Temperature, or Total Sum of Squares
VIF	Variance Inflation Factor
WWV	Warm Water Volume

1. Introduction

Rainfall, a parameter highly variable in both space and time, is difficult to forecast. Rainfall forecasts are disseminated at different time scales like short, medium and long range. The forecasts made up to a time period of 48 hours is known as short range forecasts whereas a medium range forecast is made for a period greater than three and less than seven days. These forecasts are nowadays dealt with the numerical weather prediction models. But the forecasts given for a period greater than seven days in advance; known as long-term or long-range forecasts mainly use the statistical approaches rather than numerical models. This is because of the reduced skill of the numerical models in simulating the climate over a region. As the long-term forecasts can be either quantitative or qualitative and at different time/spatial fields, a large number of statistical methods are presently in use. Thus, there are many forecast techniques in use. This chapter explains the various empirical approaches used in the linear and non-linear regression, probabilistic forecasts and a few forecast tools used at a spatial field. This consists of simple regression, multiple regression, autoregressive moving average (ARIMA) models, multiple power regression, Projection Pursuit Regression, Neural Network, parametric model, Probabilistic forecast from ensembles, Principal component analysis, Principal Component regression and Canonical Correlation Analysis. This chapter also dicusses the various measures used in the evaluation and correction of forecasts, the challenges and problems in issuing long term forecasts, various biases, limitations and common practices to minimize errors.

1.1. Historical Perspectives

Forecasting of rainfall was started in 1877 relating to monsoon rainfall over India and neighbouring regions. The first attempt to forecast the monsoon rainfall was done by H.F. Blanford (1884) based on his hypothesis that the 'varying extent and thickness of the Himalayan snow exercise a great and prolonged influence on the climate conditions and weather of the plains of northwest India'. And the operational long-range forecast (LRF) of monsoon rainfall covering the whole of India and Burma was started in 1886, followed by his cautious forecasts during 1882-85. Sir John Elliot who succeeded Blanford in 1895 utilised the weather conditions over the whole of India and surrounding regions. He used the (1) Himalayan snow cover from October to May, (2) 'local peculiarities' of pre-monsoon weather in India, and (3) 'local peculiarities' over Indian Ocean and Australia (Thapliyal, 1987). Later, Sir Gilbert Walker (1923, 1924) conducted extensive study on the variations of pressure, temperature, rainfall among other parameters and identified two pressure patterns: (1) North Atlantic Oscillation, (NAO) and North Pacific Oscillation, (NPO) in the Northern Hemisphere with a regional nature and a global scale phenomenon, and (2) Southern Oscillation (SO) in the Southern Hemisphere. The SO was later linked to the oceanic phenomenon of El Nino in the east-equatorial Pacific characterized by warming of the sea surface along the Peru coast. He was the first to introduce an objective method in forecasting by using the correlation analysis for finding the predictors in LRF. Walker (1924) also gave forecast for three subregions over India. After his work, little progress was made in this field till the 1980s. After that, there has been considerable development in the field of empirical forecasting (Gowariker, 1991; Thapliyal, 1982; Rajeevan et al, 2000)

In this chapter, various models and methods developed to forecast rainfall are discussed in the subsequent sections. The techniques, tools and models used in forecasting are also described with examples.

1.2. Types in Forecasting

Forecasting can be quantitative or qualitative based on the availability of data. It is possible to give a quantitative forecast only when the influencing factors can be quantified. The quantified information in the form of predictors will be used to give the quantitative forecasts of the predictand. The quantitative information available in the past will only be useful when the relationship between the predictors and predictand is significant and continues into the future. But when the factors influencing the predictand cannot be quantified and there is qualitative information available, then qualitative forecasts are chosen. Qualitative forecasts are more useful in planning.

1.3. Steps in Forecasting

The basic step in any kind of forecasting is to understand the problem. The main problem in rainfall forecasting is the uncertainty in the amount to be received, as the processes in producing rainfall are highly complex. The different requirement of various groups (e.g. farmers or governmental planning) also adds to the forecast issues as the level of accuracy, forecast type and the span over which these forecasts are delivered varies from group to group. Once the problem is identified, the related information needs to be collected. For rainfall, these can be different atmospheric/oceanic parameters or the processes that change the quantity. Many indices have been used in previous studies, many established relations with the quantity and other parameters need to be considered. As there will be large amount of information, it will be difficult to sort out the required and significant one. Therefore, this requires the use of some statistical tools such as correlation; one of the technique commonly used in finding the linear relation between indices. This technique is useful for parameters related linearly and if non-linear relations are involved some other techniques need to be used. The indices with high correlation and physical significance are selected and are represented as scatter plots. Such plots will show whether more than one selected index gives the same information. Then similar ones can be rejected retaining the most important ones, outliers present and trend present in the data are also found. Also the linear/non-linear relation can be identified. Once the indices are selected, they are used to formulate the model. This includes all the linear and non-linear models. When a model is developed successfully, it can be used to deliver forecasts. The forecasts should be validated with observed data. The performance of the model can be evaluated using certain statistics, which will show whether the model is good for future forecasting or the model requires certain improvement.

2. Mathematical and Statistical Methods in Forecasting

This section describes various mathematical and statistical methods that have been used in time series analysis, estimating errors and mathematical data transformations.

2.1. Time Series

2.1.1. Trend

Trend is a measure of the increase or decrease in the long-term pattern.

2.1.2. Seasonality

The existence of a pattern in the data during a period of time (say season or month) is conidered as seasonality.

2.1.3. Cyclic

This pattern will have a longer time interval than that of the seasonality and the time interval may vary, but the pattern can be observed in the data series.

2.1.4. Scatter Plot

Scatter plots give the relation between two datasets. The plots can be used to identify whether the selected variable has a positive/negative or linear/non-linear relationship with the forecast variable.

2.1.5. Simple Moving Average

Even though it is a simple analysis, simple moving average is very useful in understanding the time series. It is a smoother method, which eliminates the randomness in the data. As the name suggests, this method finds the average of the data series by moving over time. The time period over which the data series to be averaged is decided first and then the averages are calcuated. This step is repeated by moving forward in time by adding a new data point and dropping the oldest data point but by leaving the oldest observation in the data series under consideration. Thus a general equation for the calculation of moving average, T_t, is

$$T_t = \frac{1}{k}\sum_{j=-m}^{m} Y_{t+j} \tag{1}$$

where k is any odd order of the moving average, m (= (k-1)/2) is the the half-width, which denotes the number of points on either side which are included in the average, Y is the observation and t is the time at which the moving average is centered.

2.2. Basic Quantitative Statistics

2.2.1. Mean ($\overline{Y}$)

Mean is the average of all the observations, calculated as the sum of all observations divided by the number of observations. It is given as

$$\overline{Y} = \frac{1}{n}\sum_{i=1}^{n} Y_i \tag{2}$$

where Y_i denotes each observation and n is the number of observations. $\overline{Y}$ is the mean.

2.2.2. Median

Median is the middle observation when the data series is arranged in ascending order.

2.2.3. Mean Squared Deviation (MSD)

This quantity is calculated by finding the mean of the squared deviation.

$$\text{MSD} = \frac{1}{n}\sum_{i=1}^{n} (Y_i - \overline{Y})^2 \tag{3}$$

2.2.4. Standard Deviation (σ)

It gives the measure of the spread of the data and is defined as the square root of sum of squared deviations divided by (n-1) where n is the total number of observations, and (n-1) represents the degrees of freedom.

$$\sigma = \sqrt{\sigma^2} = \sqrt{\frac{1}{n-1}\sum_{i=1}^{n}(Y_i - \overline{Y})^2} \tag{4}$$

2.2.5. Variance (σ^2)

This is just the square of standard deviation.

$$\sigma^2 = \frac{1}{n-1}\sum_{i=1}^{n}(Y_i - \overline{Y})^2 \tag{5}$$

2.2.6. Covariance (Cov_{xy})

This quantity also measures the variation of a data. This is given by

$$Cov_{xy} = \frac{1}{n-1}\sum_{i=1}^{n}(X_i - \overline{X})(Y_i - \overline{Y}) \tag{6}$$

where X and Y are the two variables whose mean is given by $\overline{X}$ and $\overline{Y}$. Covariance measures how two variables change together.

2.2.7. Correlation Coefficient (r_{xy})

The correlation coefficient measures the relationship between two variables and varies between –1 and +1. It is given as

$$r_{xy} = Cov_{xy} / \sigma_x \sigma_y = \frac{\sum_{i=1}^{n}(X_i - \overline{X})(Y_i - \overline{Y})}{\sqrt{\sum_{i=1}^{n}(X_i - \overline{X})^2}\sqrt{\sum_{i=1}^{n}(Y_i - \overline{Y})^2}} \tag{7}$$

i.e., correlation is calculated as the ratio of covariance of two variables to that of respective standard deviations of variables. Therefore the data series would be standardised. A positive value indicates that the variables under consideration have a positive relation, i.e. as one increases the other also increase. A negative value indicates that the variables under consideration have an opposite relation, i.e. as one increases the other decreases.

2.2.8. Autocorrelation (r_j)

As the word indicates it is the correlation with the same data series. But the correlation is calculated with the lagged data series. It is given by;

$$r_j = \frac{\sum_{t=j+1}^{n}(Y_t - \overline{Y})(Y_{t-j} - \overline{Y})}{\sum_{t=1}^{n}(Y_t - \overline{Y})^2} \qquad (8)$$

where Y_t represents the observation at time t and j stands for the time lag.

2.2.9. Autocorrelation Function (ACF)

The pattern of autocorrelations for lag1, lag2,..., is known as autocorrelation function (ACF). It is used to identify whether or not seasonality is present in a given time series. A plot of autocorrelations against the lag is known as correlogram.

2.2.10. Least Square Estimate

The basic aim while framing a model is to fit some linear or non-linear form to a data series in such a way as to reduce the errors to a minimum. Least square is one such estimate used to minimize the errors while developing a statistical model. As the name indicates it looks for the least of the squared errors. The steps involved in the calculation of least square are given below.

1. The difference between actual (y) and estimated ($\hat{y}$) values are calculated. This is known as the error (e).
2. The square of these errors is calculated.
3. The squares of the errors are then added up.

This method of error reduction is usually known as sum of squared error (SSE)

$$\text{SSE} = \sum_{i=1}^{n}(y_i - \hat{y}_i)^2 = \sum_{i=1}^{n}(e_i)^2 \qquad (9)$$

where y_i is the observation and n is the number of observations. Another commonly used measure is the mean squared error (MSE), which is just the average of the squared error.

$$\text{MSE}=\frac{1}{n}\sum_{i=1}^{n}(y_i-\hat{y}_i)^2=\frac{1}{n}\sum_{i=1}^{n}(e_i)^2 \tag{10}$$

Thus a model is re-framed every time by adjusting the fit of linear or non-linear form until it gets the least of these error estimates.

2.2.11. Transformations of the Data Series

Transforms are usually applied so that the data appear closer to meet the assumptions of a statistical inference procedure that is to be applied, or to improve the interpretability or appearance of graphs. Whether transformation is to be applied or not is highly dependent on a particular analysis to be performed. Data transformations also help to visualise the patterns and spread the data more uniformily. Even if no analysis is to be carried out, transformation of datasets improves the interpretability. The commonly used transformations are the logarithm and square root transformations for positive data and multiplicative inverse (reciprocal) transformation for non-zero (both posiitve and negative) values. Power transformation is a family of transformations that include all the above said transformations as special cases. If transformed data series is represented by W and original data series represented by Y, then the family of power transformations are

$$W_t=\begin{cases}-Y_t^p & p<0\\ \log(Y_t) & p=0\\ Y_t^p & p>0\end{cases} \tag{11}$$

In eqn. (11), p=1 gives original data without any transformation, p=1/2 gives square root and p=-1 gives the negative reciprocal. The value of p can be any number if data is positive, p should be greater than zero if the dataset have zeros, if dataset have negative values then it should be adjusted before applying the power transformations.

3. Time Field Quantitative Forecasting Techniques

Quantitative forecasts are time-series forecasts based on one or more explanatory variables. The forecast might want to locate the causes of the behaviour. Some of these behaviours may be patterns or simply random variations. These methods are based on an analysis of historical data concerning the time series of specific variable of interest. Empirical time field quantitative forecasting of rainfall employs various statistical techniques. These techniques are discussed as follows.

3.1. Linear Methods

The linear regression models follow a linear relationship between the dependent and independent variables and the model has a linear form. The methods described in this chapter are (1) linear regression (LR), (2) multiple linear regression (MLR), (3) ARIMA models, and (4) Projection Pursuit Regression (PPR).

3.1.1. Linear Regression (LR)

The relation between two variables x and y, say; predictor (independent, explanatory) variable x and predictand (dependant, forecast) variable $\hat{y}$ (y-hat) when presented using a model, it is known as linear regression. It summarizes the relation between two variables and is obtained by fitting a straight line in such a way as to produce least error for predictions of y for the given observations of x. The method then determines the line by minimizing least square errors; hence the name ordinary least square regression (OLS). The method is sensitive to outliers.

For a given datasets of x and y , a straight line is fitted in such a way that the sum of the squared error is minimum. The model form is given as

$$\hat{y} = a + bx \tag{12}$$

where a is the intercept and b the slope of the fitted line. The difference of $\hat{y}$ (estimated value) from y (observed value) is the error or the residual, e, which is given as:

$$e_i = y_i - \hat{y}_i \tag{13}$$

for i=1,n observations. The complete form of regression equation is then

$$y_i = a + bx_i + e_i \tag{14}$$

Figure 1 shows the schematic diagram of error, intercept ad slope of the regression line.

Now the intercept a and the slope b has to be estimated for the sum of least square of errors. This is known as least square estimation. The sum of the squares of errors are given as

$$\sum_{i=1}^{n}(e_i)^2 = \sum_{i=1}^{n}(y_i - \hat{y}_i)^2 = \sum_{i=1}^{n}(y_i - (a + bx_i))^2 \tag{15}$$

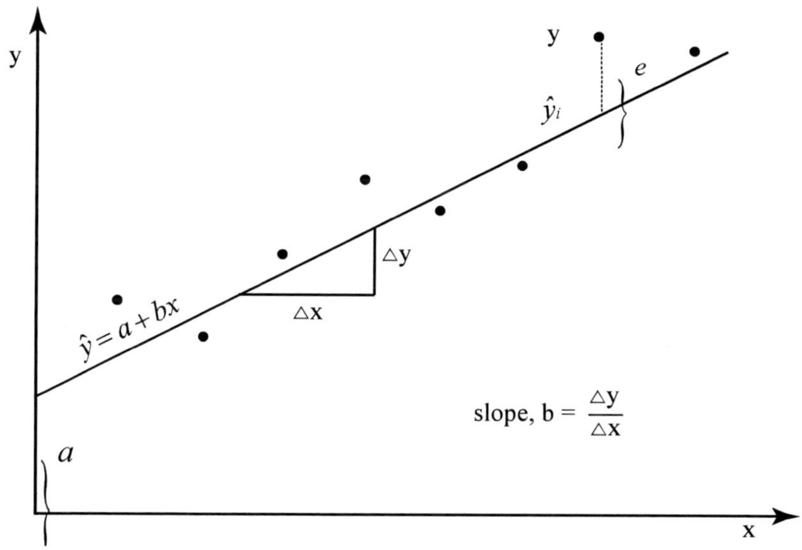

Figure 1. Schematic diagram showing simple linear regression. It shows the error, intercept and slope.

Finding the derivatives with respect to a and b (unknown variables) by substituting the value of $\hat{y}$ and setting them to zero and solving for the unknown variable, as follows.

$$\frac{\partial \sum_{i=1}^{n} (e_i)^2}{\partial a} = \frac{\partial \sum_{i=1}^{n} (y_i - (a + bx_i))^2}{\partial a} = -2\sum_{i=1}^{n} (y_i - (a + bx_i)) = 0 \quad (16)$$

$$\frac{\partial \sum_{i=1}^{n} (e_i)^2}{\partial b} = \frac{\partial \sum_{i=1}^{n} (y_i - (a + bx_i))^2}{\partial b} = -2\sum_{i=1}^{n} (y_i - (a + bx_i))x_i = 0 \quad (17)$$

Expanding eqn. 16 gives

$$2\sum_{i=1}^{n} y_i = 2na + 2b\sum_{i=1}^{n} x_i \quad (18)$$

Dividing eqn. 18 by (2n) and substituting for $\overline{y}$ and $\overline{x}$ gives the final solution;

$$a = \overline{y} - b\overline{x} \quad (19)$$

where $\overline{y} = \frac{1}{n}\sum_{i=1}^{n} y_i$ (20)

$$\overline{x} = \frac{1}{n}\sum_{i=1}^{n} x_i \quad (21)$$

Now expanding eqn. 17

$$2a\sum_{i=1}^{n} x_i + 2b\sum_{i=1}^{n} x_i^2 - 2\sum_{i=1}^{n} x_i y_i = 0 \quad (22)$$

Dividing eqn. 22 by two, and substituting for a and moving the terms of x to the left hand side (LHS) of the equation gives

$$(\sum_{i=1}^{n} x_i^2 - \overline{x}\sum_{i=1}^{n} x_i)b = \sum_{i=1}^{n} x_i y_i - \overline{y}\sum_{i=1}^{n} x_i \quad (23)$$

Substituting $\overline{y} = \frac{1}{n}\sum_{i=1}^{n} y_i$ and $\overline{x} = \frac{1}{n}\sum_{i=1}^{n} x_i$ gives

$$b = \frac{n\sum_{i=1}^{n} x_i y_i - \sum_{i=1}^{n} x_i \sum_{i=1}^{n} y_i}{n\sum_{i=1}^{n} x_i^2 - (\sum_{i=1}^{n} x_i)^2} = \frac{\sum_{i=1}^{n} (x_i - \overline{x})(y_i - \overline{y})}{\sum_{i=1}^{n} (x_i - \overline{x})^2} \tag{24}$$

3.1.1.1. Relation with Correlation Coefficient (r_{xy})

From eqn. 24, the intercept b can be written as

$$b = \frac{Cov_{xy}}{\sigma_x^2} \tag{25}$$

$$Cov_{xy} = \frac{1}{n-1}\sum_{i=1}^{n} (x_i - \overline{x})(y_i - \overline{y}) \tag{26}$$

$$\sigma_x^2 = Cov_{xx} = \frac{1}{n-1}\sum_{i=1}^{n} (x_i - \overline{x})^2 \tag{27}$$

From eqn. 7, correlation coefficient

$$r_{xy} = \frac{Cov_{xy}}{\sigma_x \sigma_y} \tag{28}$$

where σ_x and σ_y are the square roots of σ_x^2 and σ_y^2 respectively. The relation between intercept b and the correlation coefficient r_{xy} can be given as

$$b = \frac{Cov_{xy}}{\sigma_x^2} = r_{xy}\frac{\sigma_y}{\sigma_x} \tag{29}$$

3.1.1.2. Evaluating the Regression

Regression helps to determine how much of the variability in the dependent variable is explained by the independent variables. This is quantified as a sum of squares.

The total deviation to explain ($y_i - \bar{y}$) is

$$y_i - \bar{y} = (y_i - \hat{y}_i) + (\hat{y}_i - \bar{y}) \tag{30}$$

where $(y_i - \hat{y}_i)$ is the unexplained deviation e_i and $(\hat{y}_i - \bar{y})$ is the explained deviation.

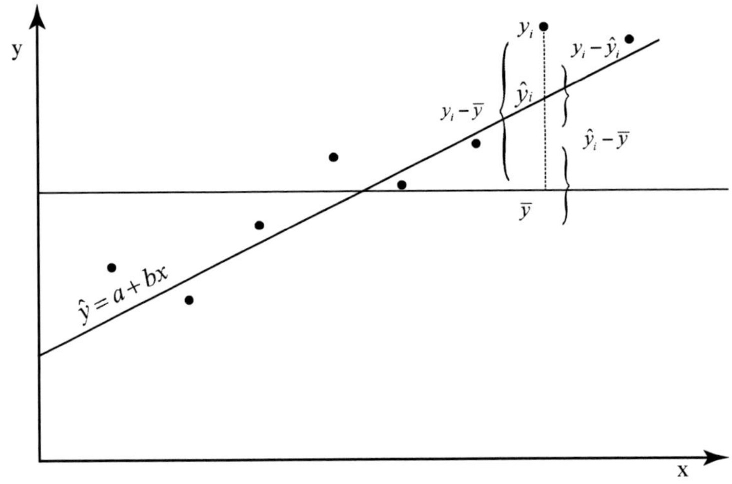

Figure 2. The contribution of unexplained and explained deviation towards Total deviation.

Squaring both sides of eqn. 30 and summing for all observations, results

$$\sum (y_i - \bar{y})^2 = \sum (y_i - \hat{y}_i)^2 + \sum (\hat{y}_i - \bar{y})^2 + 2\sum (y_i - \hat{y}_i)(\hat{y}_i - \bar{y}) \tag{31}$$

where the last term on the right hand side of the equation becomes zero. Thus

$$\sum (y_i - \bar{y})^2 = \sum (y_i - \hat{y}_i)^2 + \sum (\hat{y}_i - \bar{y})^2 \tag{32}$$

i.e. SST=SSE+SSR (33)

where SST is the total sum of squares, SSE is the unexplained sum of squares, and SSR is the explained sum of squares. Figure 2. Illustrates the total deviation.

3.1.1.3. Residuals (e)

$$SST = \sum_{i=1}^{n}(y_i - \bar{y})^2 = \sum_{i=1}^{n} y_i^2 - n\bar{y}^2 \quad (34)$$

SST is the sum of squared deviations of the y values around their mean, whereas, SSR gives the squared difference between the predictions and the sample mean of y.

$$SSR = \sum_{i=1}^{n}[\hat{y}_i - \bar{y}]^2 \quad (35)$$

SSR is related to the regression equation as given in eqn. (31) and indicates that a regression line with larger slope will have large differences from the sample mean of the predictand, hence a large SSR.

$$SSR = b^2\sum_{i=1}^{n}(x_i - \bar{x})^2 = b^2\left[\sum_{i=1}^{n} x_i^2 - n\bar{x}^2\right] \quad (36)$$

since $\hat{y}_i = a + bx_i$; $a = \bar{y} - b\bar{x}$, $(\hat{y}_i - \bar{y}) = b(x_i - \bar{x})$

SSE gives the sum of squared differences between the residuals and their mean, which is zero, or sum of squared errors.

$$SSE = \sum_{i=1}^{n}(y_i - \hat{y}_i)^2 = \sum_{i=1}^{n} e_i^2 \quad (37)$$

The estimated variance of the errors is given by

$$\sigma^2_e = \frac{1}{n-2}\sum_{i=1}^{n} e_i^{\ 2}$$

$$= \frac{1}{n-2}\{SST - SSR\} = \frac{1}{n-2}\left\{\sum_{i=1}^{n} y^2_i - n\bar{y}^2 - b^2\left[\sum_{i=1}^{n} x^2_i - n\bar{x}^2\right]\right\} \quad (38)$$

where n is the number of observations;

$$SSE=SST-SSR \quad (39)$$

3.1.1.4. Coefficient of Determination (R^2)

The correlation coefficient between forecast variable y and $\hat{y}$, the estimate (based on explanatory variable) is called the multiple correlation coefficient (R). The square of this correlation is known as coefficient of determination, R^2. R^2 is thus the squared correlation between the forecast y and the estimate $\hat{y}$. Therefore, it is defined as:

$$R^2 = r_{y\hat{y}}^{\ 2} = \frac{\sum_{i=1}^{n}(\hat{y}_i - \bar{y})^2}{\sum_{i=1}^{n}(y_i - \bar{y})^2} \quad (40)$$

R^2 is also related to SSR, SSE and SST as follows:

$$R^2 = \frac{SSR}{SST} = 1 - \frac{SSE}{SST} \quad (41)$$

For a perfect regression, SST=SSR and SSE=0, therefore R^2=1.

3.1.1.5. F-test

The F-test is a statistical test to test the significance of the regression model.

$$F = \frac{\text{explained MS}}{\text{unexplained MS}} = \frac{\text{explained SS/explained df}}{\text{unexplained SS/unexplained df}}$$

$$= \frac{\sum(\hat{y}_i - \bar{y})^2/(m-1)}{\sum(y_i - \hat{y}_i)^2/(n-m)} \quad (42)$$

where MS is the mean square, SS is the sum of squares, df is the degrees of freedom, n is the number of observations and m, the number of coefficients in the regression equation. By eqn. 40 F is also related to R^2 as:

$$F = \frac{R^2/(m-1)}{(1-R^2)/(n-m)} \quad (43)$$

The regression will explain a large proportion of the variance, if the slope coefficient is significantly different from zero (if the slope is zero, then there is no relation between the regressed parameters y and x) and then the F statistics will be large.

The P value is another measure, which gives the probability of obtaining an F statistic as large as one calculated for the data if the slope coefficient is zero. A small P value indicates that the regression is significant. By knowing the degrees of freedom for the numerator and denominator in eqn. 43, P value can be obtained from F table. The Analysis of Variance (ANOVA) for the simple regression analysis is given in Table 1.

Table 1. Analysis of Variance (ANOVA) table for simple linear regression

Source	df	SS	MS	F	P
Regression Error	m-1 n-m	SSR SSE	MSR=SSR/m-1 MSE=SSE/n-m	MSR/MSE	P- value
Total	n-1	SST=SSR+SSE			

3.1.1.6. Confidence Intervals of Regression Coefficients

The regression coefficients (a, b) calculated varies from sample to sample. This allows the construction of confidence intervals for the population. The sampling distribution of both the intercept and the slope have normal distribution with mean, α, and standard error, s.e.(a), and s.e.(b), respectively:

$$s.e.(a) = \sigma_\varepsilon \sqrt{\frac{1}{n} + \frac{\overline{x}^2}{\sum_{i=1}^{n}(x - \overline{x})^2}} \quad (44)$$

$$s.e.(b) = \sigma_\varepsilon \sqrt{\frac{\overline{x}^2}{\sum_{i=1}^{n}(x_i - \overline{x})^2}} \quad (45)$$

where σ_ε is the satandard deviation of the errors, which is unknow, but can be estimated as

$$\sigma_\varepsilon = \sqrt{\frac{\sum_{i=1}^{n}(y_i - \hat{y})^2}{n-2}} \quad (46)$$

Standard errors give an estimate of the fluctuation of a and b from sample to sample. The confidence intervals for a and b are given by

$$\alpha : a \pm t * s.e.(a) \quad (47)$$

$$\beta : b \pm t * s.e.(b) \quad (48)$$

where t^*, a multiplicative factor depends on the number of observations and the required level of confidence. α and $\beta$are the estimated confidence intervals for the coefficients a and b respectively.

3.1.1.7. t-test

The t-test is a test to determine whether the coefficients a and b are significantly different from zero. The statistic t_a and t_b are given as:

$$t_a = a/s.e.(a) \quad (49)$$

$$t_b = b/s.e.(b) \quad (50)$$

A large value of t_b indicates that the slope is significantly different from zero. A P value along with the t statistic indicates the probability of obtaining a value of t as large as the one calculated for the data, if the parameter equals zero. Small P value indicates the estimated parameter is significantly different from zero.

3.1.1.8. Assessing Performance of Regression Model

The performance of the model have been evaluated by using Correlation Coefficient (r_{xy}), Root Mean Square Error (RMSE), Absolute Error (ABSE) and BIAS for (1) training period (see Section 3.1.2.7), (2) test period (see Section 3.1.2.7). A detailed discussion and the equations used to calculate these measures are dealt under Section 6.

The accuracy of the model forecast for the training period was also compared with the model based on climatology (average of rainfall for the training period). Then, the residuals of the model are checked for any patterns. This is done by scatter plot analysis and Durbin Watson (DW) analysis of the residuals. The DW analysis is explained in Section 3.1.2.11.

3.1.2. Multiple Linear Regression (MLR)

When a dependent variable is influenced by more than one independent variable, the multiple regression method is employed. The main task in the development of the multiple regression model is the selection of useful predictors based on statistical significance and physical linkage. The various steps involved in the selection and development of the model are described as follows.

3.1.2.1. Physical Linkage and Formation of Indices

Table 2. Details of predictors used for the stepwise regression analysis for the period 1961-1994. (Adapted from Nayagam et al 2007)

No.	Parameter	Label	Level	Month	r_{xy}	Area
1	Specific Humidity (X_1)	SH300No	300mb	Nov (-1)*	0.642	80E-120E, 30-40S
2	Sensible heat flux (X_2)	SHFAu	Surface	Aug (-1)	0.640	60-80E, 60-70N
3	Meridional wind	V850Ja	850mb	Jan	0.645	0-20E, 20-50S
4	Relative Humidity (X_3)	RH600Au	600mb	Aug (-1)	0.612	30-50E, 45-60N

Table 2. (Continued).

No.	Parameter	Label	Level	Month	r_{xy}	Area
5	Meridional wind	V925Ja	925mb	Jan	0.612	0-20E, 20-50S
6	Zonal Wind (X_4)	U70No	70mb	Nov (-1)	0.590	60-120W, 30-50S
7	Meridional wind	V700Ja	700mb	Jan	0.626	0-20E, 20-50S
8	Zonal Wind	U50No	50mb	Nov (-1)	0.562	60-120W, 30-50S
9	Meridional wind	V20No	20mb	Nov (-1)	-0.596	30-90W, 30-60S
10	Meridional wind	V30No	30mb	Nov (-1)	-0.606	30-90W, 30-60S
11	Meridional wind (X_5)	V925Oc	925mb	Oct (-1)	0.556	10-40E, 30-55N
12	Zonal Wind (X_6)	U10Ma	10mb	Mar	0.519	0-60E, 35-50N
13	Zonal Wind	U20Ma	20mb	Mar	0.505	20-70E, 30-55N
14	Zonal Wind	U30Ma	30mb	Mar	0.510	20-70E, 30-55N
15	Sea Surface Temperature	SSTAu	Surface	Aug (-1)	0.503	130E-150W, 5-30N
16	Sea Surface Temperature	SSTJu1	Surface	Jun (-1)	0.454	10-50W, 30-40S
17	Geopotential Height (X_7)	Ht 250Ju	250mb	Jun (-1)	0.458	170E-120W, 10-25S
18	Sea Surface Temperature	SSTJu2	surface	Jun (-1)	0.354	65-80E, 5-20N

* (-1) denotes the previous year.

The selection of predictors can be based on statistical test and physical relation between a dependent and an independent variable. Physically meaningful predictors are identified from studies on different parameters and the dependent variable. Previously identified predictors are utilized or new predictors are made. Predictors are identified based on the physical or statistical relationship. The predictors from both regional and global fields are included.

3.1.2.2. Selecting Candidate Predictor Set

Other than the indices made from physical relationship, indices are also made based on statistical analysis. Correlation is a common method used to identify suitable predictors. Correlation gives the linear relationship between the dependent and independent variables. Therefore the regions of high correlation are identified and tested the significance. The areas that bear correlation coefficients at 1% level of significance were identified and selected for calculating indices by taking area averages of the parameter over the respective significant area. As an example (NRR7), Figure 3a & 3b of Nayagam et al 2007, shows the correlation maps of the seven selected variables having significant correlation with Kerala Summer Monsoon Rainfall (KSMR), a meteorological subdivision of India. The shaded areas have correlation significant at 1% and the selected areas are marked (rectangles in Figures 3a & 3b). Indices are made by taking area-average of the variable over the selected area (shaded area) and a 21-year sliding window correlation (Bell, 1977) between the independent variables (indices) and dependent variable were carried out. The results of the sliding correlation was checked for consistency for the entire period of analysis. Indices that have significant correlation at 5% level in all the 21-year sliding windows were retained (Figure 4). Thus, eighteen predictors were selected, which is shown in Table 2.

Stepwise regression analysis was employed to reduce the number of these indices (Draper and Smith, 1981). Thus the extensive basic predictor set is reduced to a candidate set by eliminating those with less influence on the variance of rainfall.

3.1.2.3. Procedure for Selecting the Variables for Regression

The different methods employed in selecting the predictors for the regression model are described as follows.

(1) Best Subset Regression Model

A model that fits the data well will have a high value of R^2. Hence, a model that gives maximum R^2 can be considered the best model. The value of R^2 increases as the number of independent variables in the model increase, but R^2 does not account for the degrees of freedom (df). This is solved by utilizing a quantity called an adjusted R^2. This is defined as

$$\overline{R}^2 = 1 - \left(1 - R^2\right)\frac{\text{total df}}{\text{error df}} = 1 - (1 - R^2)\frac{n - 1}{n - k - 1} \tag{51}$$

The value of $\overline{R}^2$ is compared for several models and the one with maximum $\overline{R}^2$ value is selected. But the difficulty is that a very large number of combinations of regression models is possible and it would not be wise to find $\overline{R}^2$ for all possible models. For such a case stepwise regression is the solution.

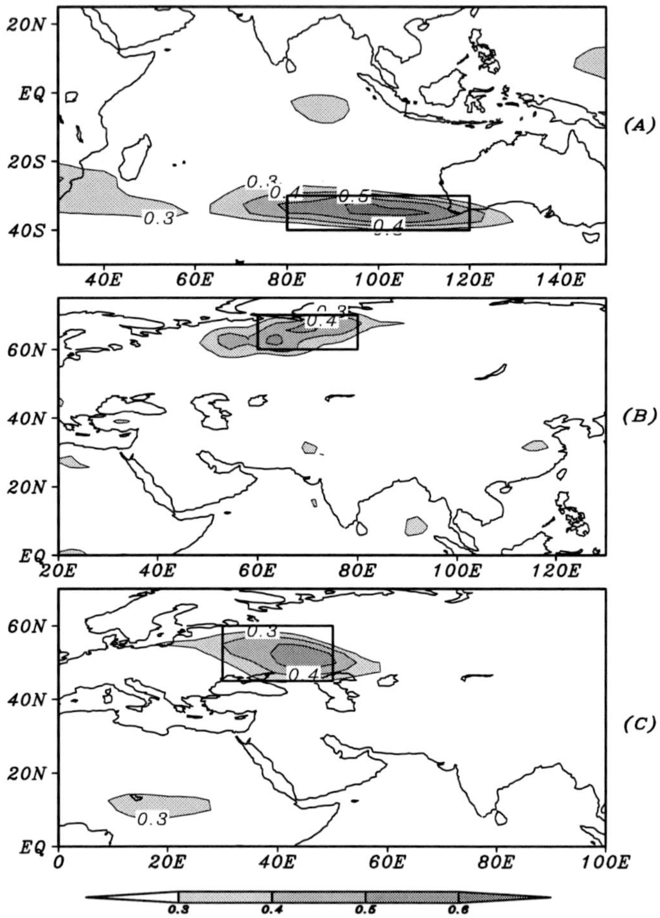

Figure 3a. Spatial Correlation of Kerala Summer Monsoon Rainfall (KSMR) with predictor parameters.(A) Specific Humidity of previous November at 300 hPa, (B) Surface Net Heat Flux, (C) Relative Humidity of previous August at 600 hPa. Shaded areas have CCs significant at 1% level. Box indicates the areas selected for making indices. Others do not satisfy the criteria (Adapted from Nayagam et al 2007).

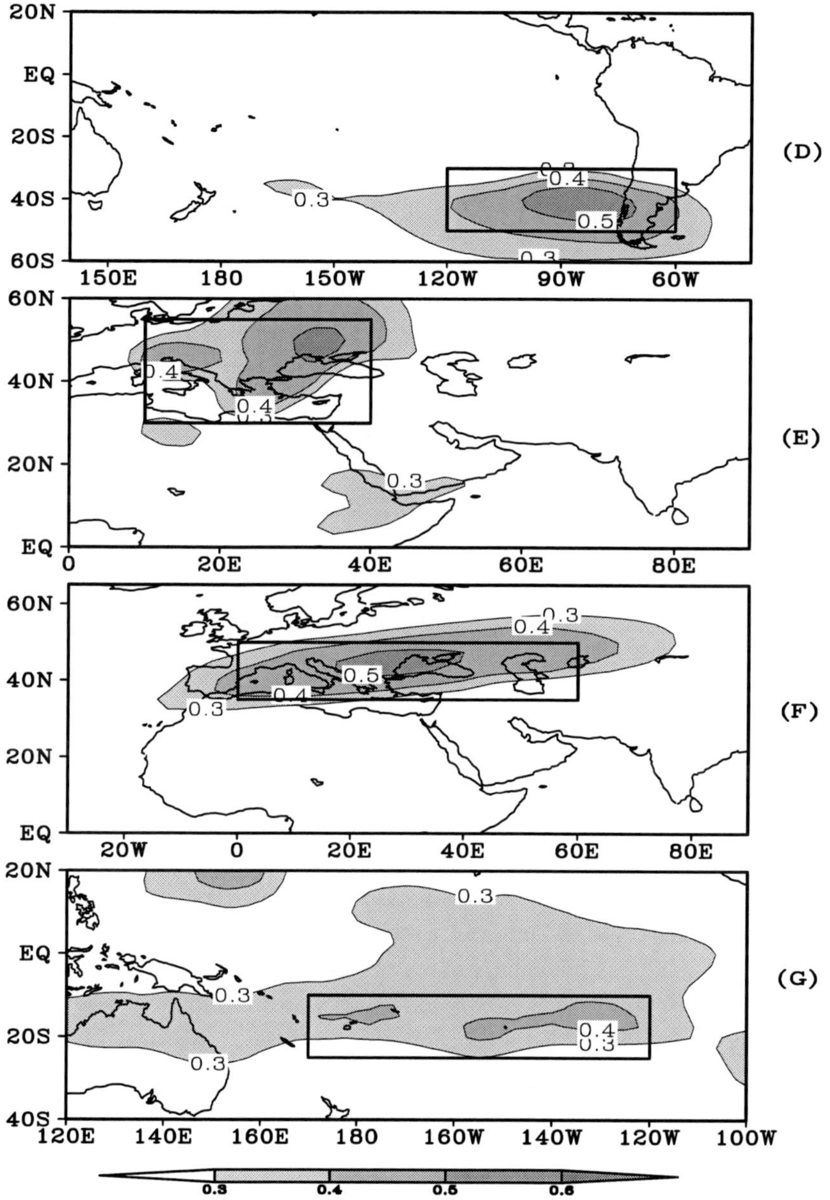

Figure 3b. Same as Figure 5a, but for (D) Zonal wind of previous November at 70 hPa, (E) Meridional wind of previous October at 925 hPa, (F) Zonal wind of March at 10 hPa, (G) Geopotential Height of previous June at 250 hPa. (Adapted from Nayagam et al 2007). (Adapted from Nayagam et al 2006).

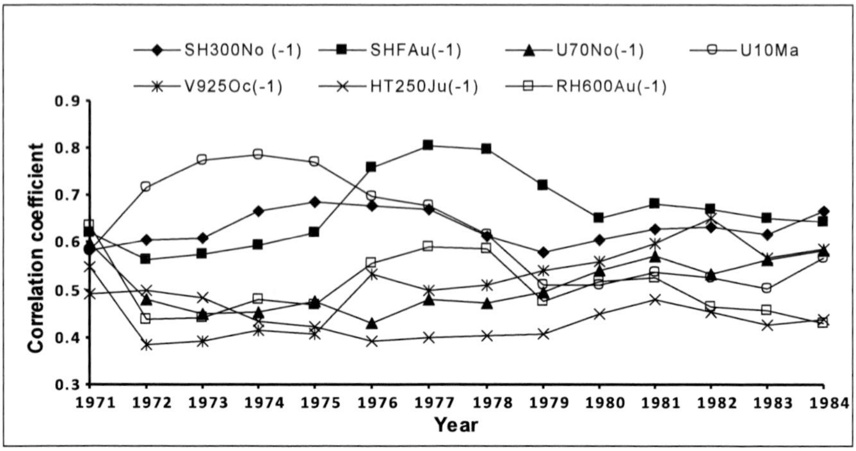

Figure 4. Twenty-one-year sliding window correlations of the seven parameters in the model with Kerala Summer Monsoon Rainfall for the period 1961-1994. Central year of the 21-year sliding window is shown. The variable details are given in Table 2. (Adapted from Nayagam et al 2007).

(2) Stepwise Regression

(a) Stepwise forward regression: In this procedure the first independent variable included has the highest coefficient of partial determination. The variable with highest coefficient of partial determination will also have the highest partial F value. It is usual to set a constant value F_{in} such that the partial F value of the best candidate variable should be greater or equal to F_{in} if that variable is to be included. Forward selection proceeds until no candidate variable qualifies to enter or all the candidates have entered the regression.

(b) Stepwise backward regression: Backward elimination starts with all the independent variables and are eliminated one at a time. In each step the variable with the least partial F value is eliminated from the regression, provided the partial F value is less than F_{out}

(c) Stepwise forward-with-a backward-look regression: This procedure is an improved method of forward selection. In this procedure the variables already in the equation are reevaluated at each stage. Due to the intercorrelation, a variable that was important at an earlier stage may not be important at a later one. Before a variable is added, the variable already in the regression with the lowest partial F value is dropped if this value is less than $F_{out.}$ When there is no improvement in the $\overline{R}^2$ value the process stops.

The example, NRR7, dealt here utilized a stepwise forward-with-a backward-look regression and from a set 18 candidate predictors, a final set of seven independent variables are chosen.

3.1.2.4. Multicolliearity

The inter-correlation between the predictors leads to multi-collinearity, i.e. the parameters are non-orthogonal. Thus a multiple linear regression model lacks in its accuracy and may lead to unclear interpretation of the regression coefficients as measures of original effects (McCuen, 1985). It imposes the problem of redundancy and unnecessary loss of degrees of freedom when they are used in large numbers (Krishnakumar et al, 1995). Moreover the coefficients would be unstable in the sense that small data revisions could have a disproportionate effect on the calculated coefficients and thus the prediction. In order to reduce this and to increase the reliability of these parameters in the prediction, the multi-collinearity is studied by the Variance Inflation Factor (VIF) method (Fox, 1991) (Table 3). It is defined as:

$$\text{VIF }(a_i) = \frac{1}{1-R_i^2} \tag{52}$$

where R_i^2 is the unadjusted R^2 when X_i is regressed against all the other explanatory variables in the model. The VIF measures how much the variance of the estimated regression coefficients are inflated compared to situations when the independent variables are uncorrelated. Values in excess of ten could significantly affect the stability of the regression coefficients (Neter et al., 1990). If the independent variables are significantly correlated then a Principal Component Regression Method (discussed in Section 5.2) can be utilised. Table 3 gives the VIF calculated for each of the independent variables in the given example, NRR7. The values are all less than 1.7 indicating that there is no multi-collinearity in the model.

The relationship between each pair of variables can be visualized on scatter-plot matrix. Each column shows the relationship between the parameter listed in that column with the other seven parameters named in the respective rows. Figure 5. shows the scatter plot matrix for the seven selected predictors. The correlation coefficients between the dependent variable and seven independent variables are listed atop the columns. From the figure it can be seen that the KSMR is positively related to all the parameters while the independent parameters do not show much relation among themselves though feeble relations exist. The plot does not exhibit any noticeable pattern and this implies that the predictors are practically independent. So the interdependencies among the predictors, if any, are well below the potential to affect the forecast.

Table 3. Results of Variance Inflation Factor and the t-test for the selected variables. (Adapted from Nayagam et al 2007)

No. as in Table 2	Variables	Std. Error	t value	P value	VIF
1	Specific Humidity	1586.870	1.790	0.085	1.677
2	Sensible Heat Flux	4.670	2.953	0.007	1.496
4	Relative Humidity	7.897	2.195	0.037	1.623
6	Zonal Wind 70mb	11.055	3.629	0.001	1.256
11	Meridional Wind	24.408	3.118	0.004	1.306
12	Zonal Wind 10mb	4.103	3.798	0.001	1.292
17	Geopotential Height	1.523	2.849	0.008	1.194

3.1.2.5. Development of Regression Model

From among these refined candidate predictors, various regression models were formulated with the following general form using different iteration schemes.

$$\hat{y}_i = b_0 + b_1 x_{1,i} + b_2 x_{2,i} + \cdots + b_n x_{n,i} \tag{53}$$

or

$$y_i = b_0 + b_1 x_{1,i} + b_2 x_{2,i} + \cdots + b_n x_{n,i} + e_i \tag{54}$$

where $e_i = y_i - \hat{y}_i$; i=1,....,k

In eqns. (52) and (53), each of the n predictor variables has its own coefficient analogous to the slope b in the linear regression equation (eqn. 24). When n = 1 the multiple linear regression reduces to the simple linear regression and b_0 is equivalent to intercept a in the linear regression equation.

The method of least squares is then used to determine b_0, b_1, b_2, etc so that the sum of squares of the error term is minimized.

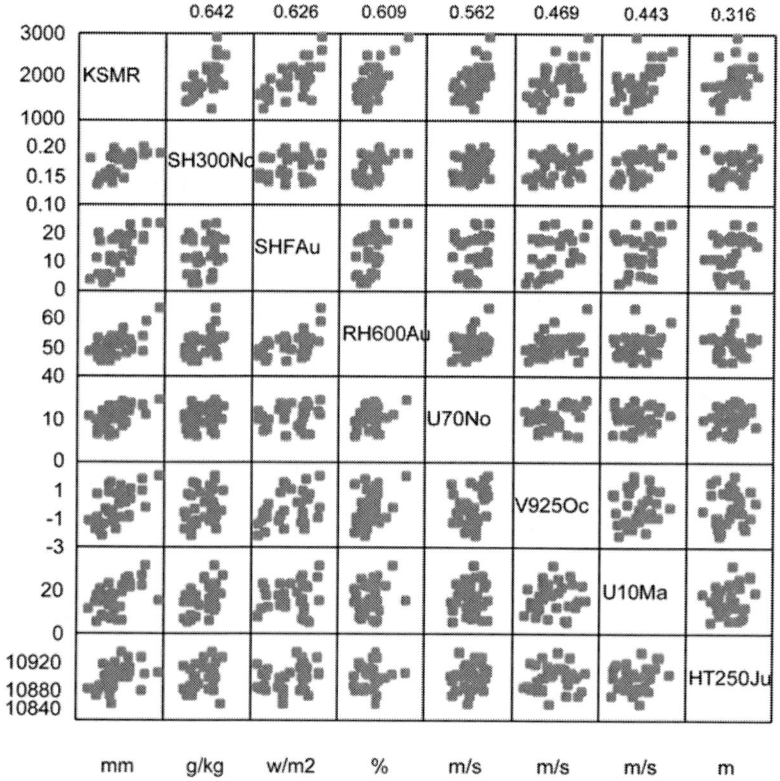

Figure 5. Scatter plot matrix of correlation between KSMR and predictor variables. The variable on the vertical axis is the variable named in that row; the variable on the horizontal axis is the variable named in that column. The variable details are given in Table 2. Values on top of the columns represent the respective correlation coefficients of KSMR with each predictor. Bottom labels represent the units of the variables. (Adapted from Nayagam et al 2007)

3.1.2.6. Regression Coefficients

The regression coefficients b_0, b_1, b_2,…..,b_n are determined by minimizing the square of the errors (i.e. min. $\sum e_i^2$). As in simple regression, the coefficients are solved by taking the partial derivatives of the error term with respect to each of the coefficients and setting their derivatives to zero. The equations are solved to find the unknowns to get the regression coefficients. In the given example, NRR7, the final equation obtained with the seven selected predictors is

$$Y = 2841.2(X_1)+13.8(X_2)+17.3(X_3)+40.1(X_4)+76.1(X_5)+15.6(X_6) +4.3(X_7)-47632.3 \quad (55)$$

where X's are the predictors (See Table 2) and Y the predicted rainfall. The regression coefficients in the model are not standardized and show disproportionate values. The coefficients can be compared only when it is expressed in standardized form.

The correlation between the observed and the forecast is called multiple correlation coefficient (R). For the example, NRR7, The model has a multiple correlation coefficient (R), correlation between the observed and the predicted, of 0.943 and the coefficient of determination, which is the proportion of variance explained by the explanatory variables in the regression model, of 88.89%. The model is successful in capturing the variance, thereby explaining the interannual variability of rainfall.

3.1.2.7. Training and Test Period

Usually regression models are trained for a period known as the training period. The regression coefficients are estimated for this training period following all the assumptions and tests detailed in Section 3.1.2.10. The regression coefficients thus estimated are then utilized to give forecast. These forecasts are validated with other observations. This period over which the model is validated is known as the test period.

3.1.2.8. F-test

The F-test is to test the significance of the regression model. In Section 3.1.1.2, it has been shown that the SST=SSE+SSR. In other words, the total degrees of freedom is the sum of unexplained degrees of freedom and explained degrees of freedom, i.e.

$$\text{Total df} = \text{unexplained df} + \text{explained df} \quad (56)$$

$$\text{Thus } n-1 = (n-k-1)+k \quad (57)$$

where n is the number of observations, k is the number of coefficients. Expressing eqn. 56, in terms of mean square

$$MST=SST/\text{total df} \tag{58}$$

$$MSE=SSE/\text{unexplained df} \tag{59}$$

$$MSR=SSR/\text{ explained df} \tag{60}$$

where MST, the total mean square, MSE, the error mean square and MSR, the regression mean square, df is the degrees of freedom, SSR, SSE and SST are given by eqns. 33, 34, 36.

Thus the F-statistic become

$$F = \frac{\text{explained MS}}{\text{unexplained MS}} = \frac{MSR}{MSE} = \frac{\sum_{i=1}^{n}(\hat{y}_i - \bar{y})^2 / k}{\sum_{i=1}^{n}(y_i - \hat{y}_i)^2 / (n-k-1)} \tag{61}$$

where n is the number of observations, and k is number of observations -1.

By eqn. 40, F is related to R^2 as follows:

$$F = \frac{R^2/k}{(1-R^2)/(n-k-1)} \tag{62}$$

Interpretation of F statistic and P value is similar to that for linear regression in Section 3.1.1.5. Table 4. contains the Analysis of Variance (ANOVA) for the multiple regression analysis.

Table 4. Analysis of Variance (ANOVA) table for multiple linear regression

Source	df	SS	MS	F	P
Regression Error	k n-k-1	SSR SSE	MSR=SSR/k MSE=SSE/n-k-1	MSR/MSE	p-value
Total	n-1	SST=SSR+SSE			

Adapted from Nayagam et al 2007.

3.1.2.9. t-test

The t-test when applied to an individual coefficient determines the significance of that coefficient in the presence of all other explanatory variables. The confidence intervals for each coefficient are detrmined in the presence of all other explanatory variables because multiple regression takes care of the interdependence of the other independent variable to model dependent variable. Therefore, for each coefficient of the model; the standard error, confidence intervals and its significance is calculated. Standard error is calculated as

$$t = \frac{b_i}{s.e.(b_i)} \tag{63}$$

where b_i is the regression ccoefficient and s.e. is the satndard error of the coefficient. Thus for each coefficients of regression model the standard deviation, t value and its significance (P value obatained from statistical table of t statistic) are calculated. Now the confidence intervals is calculated as $b_i \pm t^* s.e.(b_i)$ where t^* corresponds to the critical value of t statistic which depends on the number of observations n, number of regression coefficients k, and the required confidence interval. The degrees of freedom of a parameter estimate and t-test are the same.

The result of all the statistical tests described is given for the example, NRR7. In the t-test, the effect of each variable is assessed by its P value. Here the P value of all the regression coefficients are found to be small, indicating that each of the coefficients contribute significantly to the model developed (Table 3). The smallest P value is obtained for zonal wind at 70 and 10 mb, which are highly significant than all other explanatory variables in the regression model. The F-statistic, which tests the significance of regression model, gives F=29.561 and the P value in the ANOVA test (Analysis of Variance) is very small (<0.0005) implying that the explained variance is more than the unexplained variance. This strongly suggests that the included predictors collectively account for a significant part of the variance in the KSMR. The predictions obtained during the training and the test is shown in Figure 6.

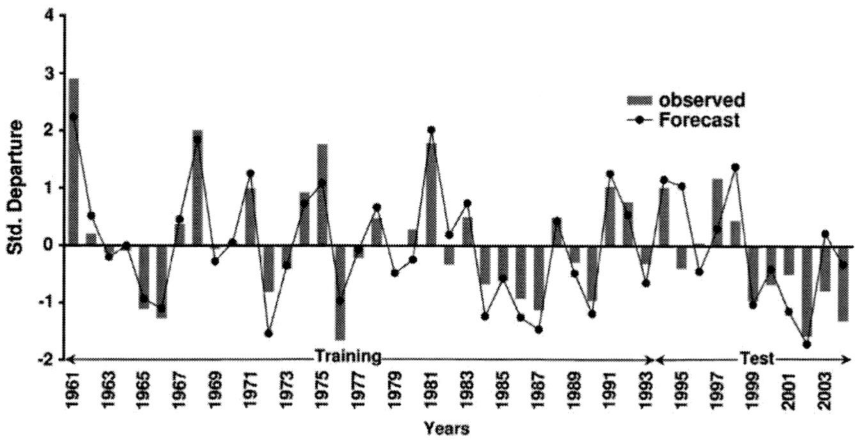

Figure 6. The forecast of KSMR for the period 1961-2008. The period 1961-1994 represents the training and1995-2008 represents the test. (Adapted from Nayagam et al 2007)

3.1.2.10. Assumptions in Multiple Linear Regression Models

A linear multiple regression model has four basic assumptions as follows:

1. Model form which refers to the form of the relationship between forecast variable and explanatory variables. If the form is incorrect, forecast may be inaccurate and the F-test, t-test and confidence intervals are not strictly valid any longer.
2. The assumption of independence of residuals is also directly related to the F, t-test, R^2 and confidence intervals.
3. Homoscedasticity means that the residuals have the same variance throughout. If this is violated, the statistical tests are not valid.
4. Normality of residuals assumes a normal distribution for the error terms.

For the example, NRR7, there are seven independent variables and the residuals of these variables after regressing them with the predictand variable shown in Figure 7. The figure shows no specific pattern and are not related to the fitted value. Hence these seven variables form the regression model.

3.1.2.11. Durbin Watson (DW)

The durbin watson tests that there is no lag1 autocorrelation present in errors obtained from forecasting. It is used to find the relationship in the residuals.

$$DW = \frac{\sum_{t=2}^{n}(e_t - e_{t-1})^2}{\sum_{t=1}^{n} e_t^2} \quad (64)$$

where e_t is the residual at the time t and e_{t-1} is the residual at time t-1. It is the ratio between the sum of the squares of the difference between successive errors to the sum of the squared errors. The DW statistic checks the significance of the assumption that the residuals for successive observations are uncorrelated. DW ranges from 0 to 4 and DW < 2, indicates that there is some positive lag1 autocorrelation present in the errors. DW > 2, indicates that there exists some negative autocorrelation and DW < 2, indicates a positive autocorrelation. If there exists any kind of significant lag1 autocorrelation, then the assumption of independence of residuals is violated and the model can be improved further (Makridakis et al, 1998).

For the given example, NRR7, the residuals were also checked to find out if there is any pattern present. Scatter plots of residuals with the predictors and fitted values were used to visualize the patterns in it. If some organized pattern is found, it means that there exists some relationship between the fitted values and the residuals so that the model can be improved further by extracting that relationship. For the example, NRR7, no pattern was found between the residuals and the fitted values and DW = 2.099, which suggests that the lag1 autocorrelation is negligibly small in the residuals.

3.1.2.12. Assessment of Model

The performance of the model is assessed by computing the RMSE (see Section 6.1.1.5), BIAS (see Section 6.1.1.1), ABSE (see Section 6.1.1.2) and also with climatological predictions. For the example, NRR7, rxy (see Section 2.2.7) is 0.943 during the model development period, RMSE = 6.60 %, BIAS = –0.26% and ABSE = 5.33% of mean rainfall. For the test period, rxy = 0.49, which is also statistically significant. For the period 1961-2008, the rxy has dropped to a value of 0.87. The forecast and observed are closely related with a correlation, significant at 1% level. The RMSE, BIAS and ABSE for the test period have increased to 15.80%, 6.20% and 13.15% of mean rainfall, respectively. In most cases the model error is less than half of the standard deviation (σ =368.58 mm) of observed rainfall. In most cases, the model error is less than half of the standard deviationof observed rainfall. Of the 44 years, only in three years the error was more than the standard deviation. The RMSE for the climatological prediction is 17.90 % of mean rainfall. The BIAS and ABSE are -5.40% and 15.16% of mean rainfall, respectively. As the model errors are less than the climatological prediction, the model is good.

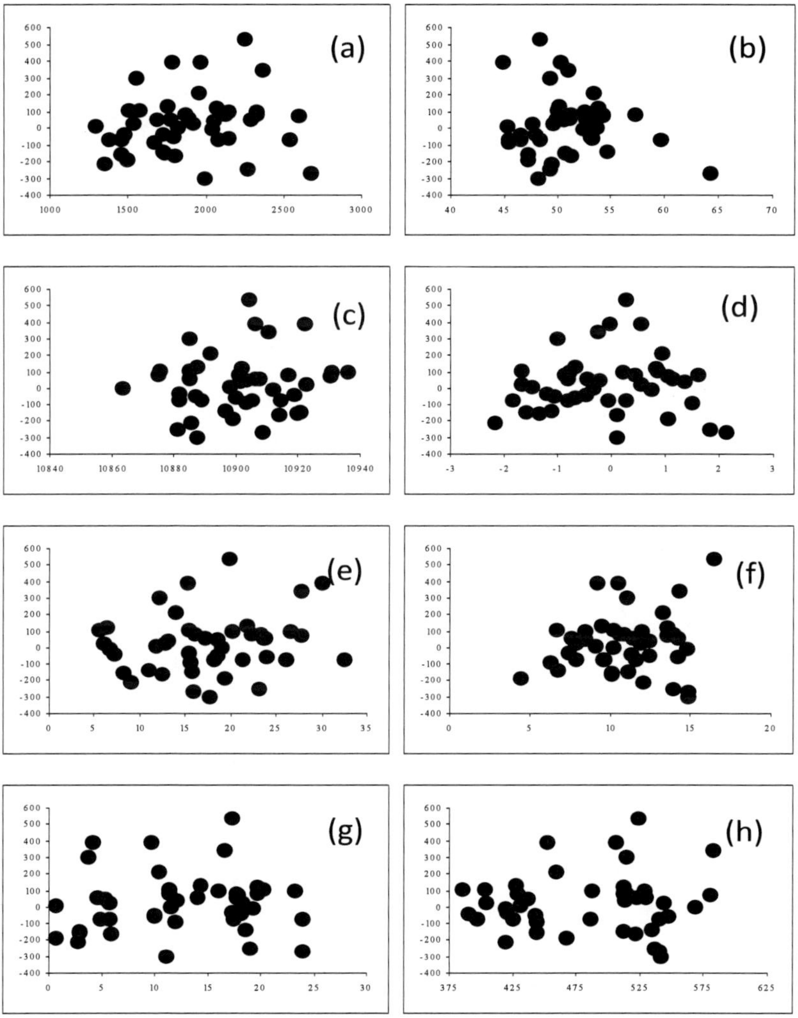

Figure 7. The Residuals plotted against (a) fitted values (b) Rhum600a (c) hgt250jn (d) V925o (e) U10mr (f) U70no (g) ShtflAu (h) Shum300no. (See Table 2 for parameter details).

3.1.3. Autoregressive Integrated Moving Average (ARIMA) Model

The Autoregressive Integrated Moving Average models (ARIMA) are widely used and being popularized by Box and Jenkins (1970). This section discusses, (1) the concepts and tools in time series analysis (2) Autoregressive (AR) model, (3) Moving Average (MA) model, (4) Autoregressive Moving Average (ARMA) model and (5) ARIMA model.

3.1.3.1. Concepts and Tools in Time Series Analysis

3.1.3.1.1. Autocorrelation and Autocorrelation Function (ACF)

As explained in Section 2.2.8, autocorrelation is the correlation of the same data series at different lags and the pattern of its corresponding correlations are known as autocorrelation function. More details on its interpretation based on its significance is discussed in Section 3.1.3.1.2

3.1.3.1.2. Interpretation of Autocorrelation Analysis

An important factor while considering the autocorrelation is to find out whether the autocorrelation is significantly different from zero. A random time series is expected to have zero for each autocorrelation, which does not hold true for a finite series. But the distribution of autocorrelation for white noise (data series with no patterns) is shown to have a normal curve with mean zero and standard error $1/\sqrt{n}$ (Anderson 1942,Barlett 1946) where n is the number of observations. Thus for a series of white noise, 95% of all sample autocorrelation coefficients should lie within $\pm1.96/\sqrt{n}$. The value 1.96 is obtained from the table of normal probabilities for standard normal distribution. If this criterion is not satisfied the series is not a white noise. Any autocorrelation value outside the $\pm1.96/\sqrt{n}$ limits indicates the presence of a pattern within the series. The $\pm1.96/\sqrt{n}$ limits are known to be the critical values and are considered while plotting ACF.

3.1.3.1.3. Partial Autocorrelation Function (PACF)

Partial autocorrelation function is used to measure the degree of association between y_t and y_{t-j} when the effects at other time lags 1, 2, 3,…., j-1 are removed. If a significant autocorrelation is present between y_t and y_{t-1} then there will be correlations between y_t and y_{t-2} because they are related to y_{t-1}. So, to measure the relationship between y_t and y_{t-2}, the effect of y_{t-1} is removed. This is what happens in partial autocorrelations. The partial autocorrelation of order j can be expressed in the form of a regression equation as:

$$y_t = b_0 + b_1 y_{t-1} + b_2 y_{t-2} + \ldots\ldots + b_j y_{t-j} \tag{65}$$

In eqn (64), each coefficient is multiplied by the previous forecast of y_t. This has a form similar to that of autocorrelation equation and the first partial autocorrelation is equal to the first autocorrelation.

The distribution of partial autocorrelation for white noise is shown to have a normal curve with standard error $1/\sqrt{n}$ (Quenouille 1949). Therefore, the critical limits of PACF also have the same value as of ACF. Thus, $\pm 1.96/\sqrt{n}$ is considered while plotting PACF.

3.1.3.1.4. Portmanteau Tests

A portmanteau test determines whether all autocorrelation considered for the data series are significantly different from zero. One such test is the Box-Pierce test, which is based on Box-Pierce Q statistic given by

$$Q = n\sum_{j=1}^{h} r_j^2 \tag{66}$$

where n is the number of observations and h is the maximum lag considered. Usually a value of h = 20 is used. Large values of positive or negative autocorrelation give large values of Q. This statistic is used to analyse the model residuals and has a Chi-square distribution when the residuals are white noise. Q follows Chi-square with (h-m) degrees of freedom where m is the number of parameters estimated in the model. A modified Q statistic is the Ljung-box statistic, which is given by

$$Q = n(n+2)\sum_{j=1}^{h} \frac{1}{n-j} r_j^2 \tag{67}$$

If the residuals are white noise, they should be uncorrelated, insignificant and Q follows the Chi-square distribution and the statistic exhibits a small value.

3.1.3.1.5. Seasonality

Seasonality is defined as a pattern that repeats itself over fixed interval of time (Section 2.1.2). It is identified by the presence of a large autocorrelation coefficient or large partial autocorrelation coefficient at a seasonal lag. Often, autocorrelation at multiples of the seasonal lag will also be significant, such as lag 24 or lag 36 (i.e. lag by 24 or 36 months, repsectively).

3.1.3.1.6. Stationarity and Non-stationarity of Time Series Data

When the data fluctuate along a constant mean and have constant variance over time then the data is said to be stationary. The autocorrelation of stationary data approaches zero at lesser number of time lags. For a non-stationary data, the autocorrelations differ from zero and the autocorrelations decrease slowly with increasing time lags. The autocorrelations and the PACF at lag1 will be large for non-stationary data. Therefore, the patterns present in the time series need to be removed. As such, a differencing process should be performed until an obvious pattern such as a trend or other patterns in the data fades away. The differenced time series is expressed as the change between each observation in the data series (i.e. the first differences of a non-stationary time series value y_1, y_2, y_n are described as $z_t = y_t\text{-}y_{t-1}$ where t = 2,…., n). If the differences of a time series are still not stationary, the second differences should be implemented. The second differences of time series value y_1, y_2, y_n are $z_t = y_t - 2y_{t-1} + y_{t-2}$ for t = 3, 4,..., n and thus the differenced series will have n-1 values. Non-stationary datasets with positive or negative trends can change its direction in an unpredictable manner. A model for such time series is called random walk model, whose basic form is given by

$$y_t\text{-}y_{t-1}= e_t \tag{68}$$

where e_t is white noise (Section 3.1.3.1.2). The random walk model is a case of autoregressive model when the autoregressive coefficient becomes one.

Seasonal differencing is utilized for seasonal non-stationary datasets. Similar to the simple differencing, this also takes differences between observations. It, however, takes the difference between an observation and the observation from the previous year separated by a certain time period. The time period is chosen according to the data under consideration. If the data has n observations a year then the time period is n. For example, if the data is monthly (twelve observations an year) then differencing is done between the current observation and the observation twelve months before. If there are four observations an year, then differencing is done between current and the observation four period back. The seasonal pattern present in a data series can be easily understood from its ACF and PACF plots. ACF will show large autocorrelations at seasonal intervals that will be larger than the adjacent autocorrelations and PACF exhibits large values at lags corresponding to the time period of the raw data. If some non-stationarity exists after seasonal differencing then the differenced series is subjected to first difference. The difference between first difference and seasonal difference is that first difference reflects the change between an observation and the next whereas the seasonal difference reflects the change between observations from one year to the next.

The following shows an example (MOM9) using the monthly rainfall of Amman airport station (Momani 2009). The first step is to check the stationarity of the data. This is checked by plotting the ACF and PACF of original data (Figure 8). It is clear that the data is not stationary, as there is significant autocorrelation and partial autocorrelation present in the data. Therefore, the data is subjected to seasonal first difference. The data is now stationary in mean and variance. ACF and PACF of the series after differencing is shown in Figure 9.

Methods to diagnose a non-stationary series have been discussed in Section 3.1.3.1.6. Now, How to identify a stationary series? A test known as Dickey-Fuller test is used for such purposes. It is a unit root tests whose critical value are obtained from the Dickey-Fuller table.

A regression equation as below is considered

$$y'_t=\phi y_{t-1}+b_1 y'_{t-1}+b_2 y'_{t-2}+....+b_p y'_{t-p} \qquad (69)$$

where y'_t denotes the differenced time series y_t-y_{t-1}. The value of p is usually set to three so that if y_t requires differencing the resulting ϕ will have a value close to zero, where the value of ϕ is estimated using ordinary least square methods. As the assumption of t-test is not valid for differenced series, the value of ϕ corresponding to the differenced series is compared using the Fuller table (1976) and a significant negative value of ϕ indicates a stationary time series. Further information can be found in Dickey et al (1986).

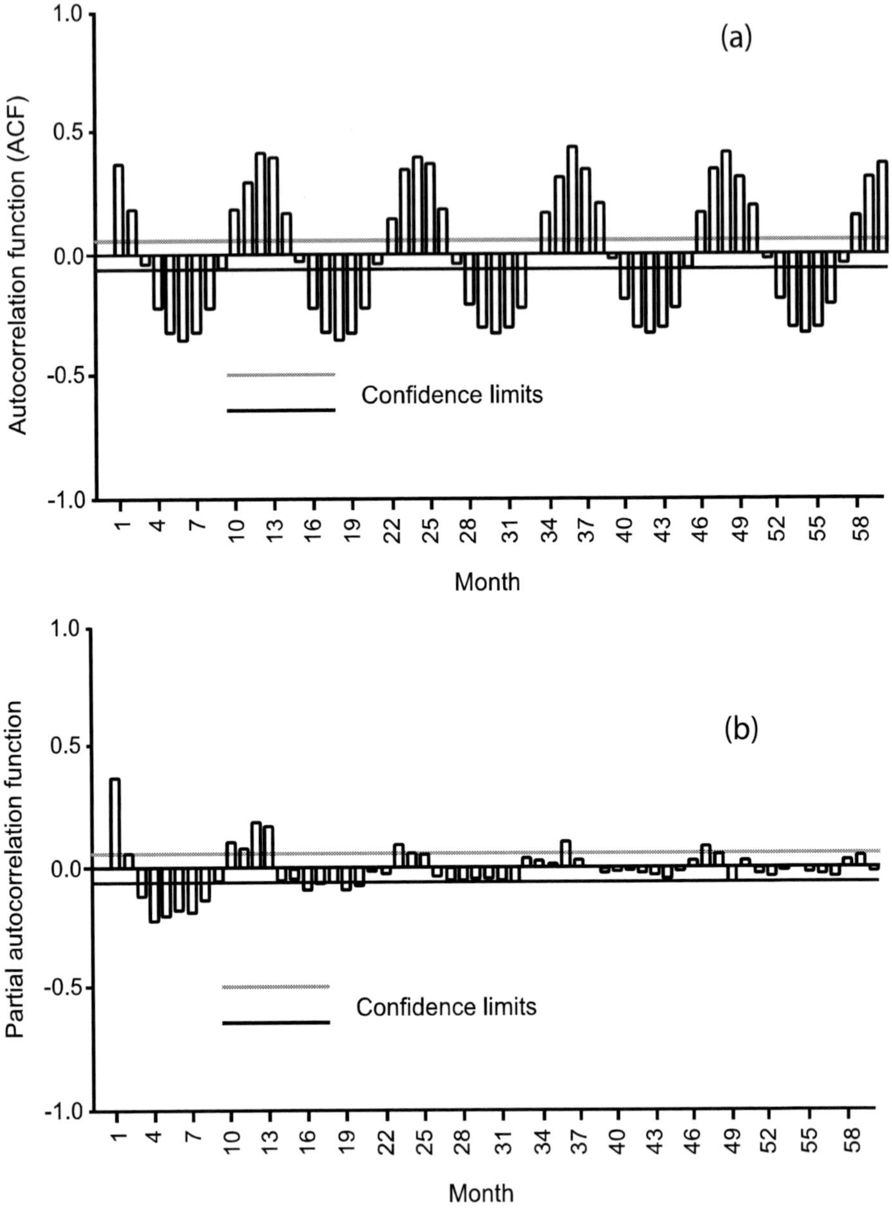

Figure 8. The plots of (a) Autocorrelation and (b) Partial Correlation Function for the monthly rainfall of Amman airport station. (Adapted from Momani 2009).

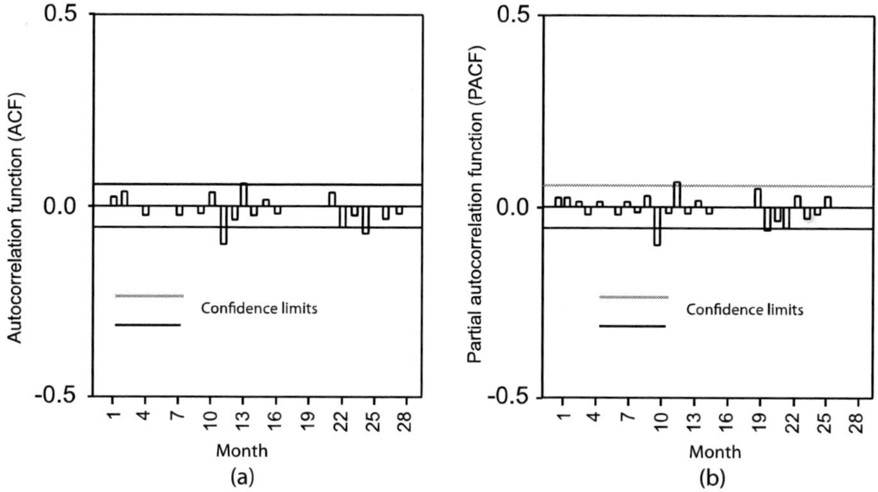

Figure 9. The plots of (a) Autocorrelation and (b) Partial Correlation Function for the first order seasonal differecing and de-seasonalised rainfall data. (Adapted from Momani 2009).

3.1.3.2. Backshift Operator

The backshift operator (B) is an operator used to shift the data by the required period. If the data need to be shifted by one, two or three periods, the equatons are

$$By_t = y_{t-1} \tag{70}$$

$$B(By_t) = B^2 y_t = y_{t-2} \tag{71}$$

$$B^2 (By_t) = B^3 y_t = y_{t-3} \tag{72}$$

Hence, the general equation for a season is

$$B^s y_t = y_{t-s}. \tag{73}$$

A differencing process is also expressed using a backshift operator. Thus a first difference is expressed as

$$z_t = y_t - y_{t-1} = y_t - By_t = (1-B)y_t \tag{74}$$

and the second difference will be $(1-B^2)y_t$, and so on. The operator is also useful in expressing a first difference of the first difference (i.e. second order difference). It is given as $(1\text{-}B)^2 y_t$. It should be noted that second difference and second order difference are different. Generally n^{th} order difference is given by $(1\text{-}B)^n y_t$. A series subjected to seasonal difference and first difference are expressed using B as

$$(1\text{-}B)(1\text{-} B^s)\, y_t = y_t - y_{t-1} - y_{t-s} + y_{t-s-1} \qquad (75)$$

3.1.3.3. Different Types of Autoregressive Models

In this section, the AR, MA, ARMA and ARIMA models are discussed. Different models discussed below can also be expressed in terms of Autogressive Integrated Moving Average (ARIMA) models. The general form of non-seasonal ARIMA model is known as ARIMA (p,d,q), where p is the order of autoregression, d is the degree of first differencing and q is the order of the moving average. Thus, a model with no AR (Section 3.1.3.3.1) , no MA (Section 3.1.3.3.2) and no differencing (Section 3.1.3.1.6) can be expressed in terms of (p,d,q) as ARIMA(0,0,0) (e.g. white noise model). Similarly, a model with only differencing is denoted as ARIMA(0,1,0) (e.g. random walk model). If the value corresponding to p, d, q is zero, then the model name can be abbreviated expressing only the valid part (e.g. ARIMA(1,0,0) can be written as AR(1) because only the AR part is present and the differencing and MA part are absent. Similarly, ARIMA(0,0,1) can be written as MA(1) as the AR and differencing part are absent. A model with differencing and MA part are given as ARIMA(0,1,1) or IMA(1,1).

3.1.3.3.1. Autoegressive (AR) Model

The autoregression models have time-lagged values of the forecast as the independent variables. It is different from ordinary regression because the assumption of independence of residuals is not followed. A basic equation of AR is

$$Y_t = b_0 + b_1 Y_{t-1} + b_2 Y_{t-2} + b_3 Y_{t-3} + \ldots\ldots + b_n Y_{t-p} + e_t \qquad (76)$$

where Y_{t-1}, Y_{t-2},.. are the time lagged values of the forecast, b_0, b_1 are the regression coefficients, and e_t is the error term. An Autoregressive model of first order can be expressed in terms of ARIMA model as ARIMA(1,0,0). ARIMA(1,0,0) or AR(1) model has no differencing term and no moving average part. Therefore only autoregression need to be considered. Thus an AR(1) model can be written as

$$y_t=c+\phi y_{t-1}+e_t \tag{77}$$

where c is the constant, ϕ the autoregressive coefficient with limits –1 and +1 and e_t the error term distributed normally with mean = 0 and variance = 1. The model reduces to a white noise model where $\phi = 0$, and becomes a random walk model with $\phi = 1$. In real-life situations, the order of the autoregression is unknown. It is, however, possible to determine from the ACF and PACF plots and thus identify the AR(1) model. For the AR(1)model, the theoretical ACF decay exponentially whereas PACF is large at lag1 and zero at other lags, but this may not be true in actual cases due to presence of errors.

A more general form of AR model of pth order can be written as

$$y_t=c+\phi_1 y_{t-1}+\phi_2 y_{t-2}+.....+\phi_p y_{t-p}+e_t \tag{78}$$

where c is the constant, ϕ_k is the kth autoregressive parameter and e_t is the error term at time *t*.

There are some requirement imposed on MA models, such as for p=1, $-1<\phi_1<1$. For p=2, $-1<\phi_2<1$, $\phi_2+\phi_1<1$ and $\phi_2-\phi_1<1$. The conditions are more complicated for higher values of p. The general model equation (eqn. 78) can also be expressed in terms of B (Section 3.1.3.2), By retaining only the constant and error term to the right hand side of the equation, eqn. 78 becomes

$$(1-\phi_1 B-\phi_2 B^2-.....-\phi_p B^p)\, y_t=c+e_t \tag{79}$$

For AR(2) model, the ACF will decay in a damped sine-wave manner, and PACF will show two non-zero partial correlations. Thus, the patterns of ACF and PACF are just the reverse of what are observed for the MA(2) model. For $q\geq 2$, the PACF equals zero for AR(p) model beyond lag p and the ACF will show an exponential decay of the damped sine-wave.

3.1.3.3.2. Moving Average (MA) Model

For a moving average model, there is a relationship among the successive errors and the model form is

$$Y_t=b_0+b_1 e_{t-1}+b_2 e_{t-2}+b_3 e_{t-3}+.....+b_n e_{t-q}+e_t \tag{80}$$

A first order MA(1) or ARIMA (0,0,1) model depends on the error terms e_t and e_{t-1}. So, the model form is

$$y_t = c + e_t - \theta_1 e_{t-1} \quad (81)$$

where $-1<\theta_1<1$ and e_t is normally distributed with mean = 0 and variance = 1. θ_1 in the eqn. 81 is negative, which is a convention for ARIMA models. The ACF has only one non-zero autocorrelation at lag 1 and the PACF decreases exponentially.

In general, a MA model of qth order can be written as

$$y_t = c + e_t - \theta_1 e_{t-1} - \theta_2 e_{t-2} - \ldots - \theta_q e_{t-q} \quad (82)$$

where c is the constant, θ_k is the kth moving average parameter and e_{t-j} is the error term at time t-j. There are some requirement imposed on MA models, such as for q=1, $-1<\theta_1<1$. For q=2, $-1<\theta_2<1$, $\theta_2+\theta_1<1$, and $\theta_2-\theta_1<1$. The conditions are more complicated for higher values of q. The general model equation (eqn. 82) can also be expressed in terms of B, the backshift as follows

$$y_t = c + (1 - \theta_1 B - \theta_2 B^2 - \ldots - \theta_q B^q)\, e_t \quad (83)$$

For MA(2) model, the ACF will show two non-zero autocorrelations, and the PACF will decay in a damped sine-wave manner. For $q \geq 2$, the PACF shows an exponential decay of the damped sine-wave, and the ACF equals zero for MA(q) model beyond lag q.

3.1.3.3.3. Autoregressive Moving Average (ARMA) Model

A combination of AR and MA models are called autoregressive moving average (ARMA) models, which can be applied only to stationary datasets.

A combination AR (1) and MA (1) model can be expressed as ARMA (1,1) or ARIMA (1,0,1) whose basic model equation is

$$y_t = c + \phi_1 y_{t-1} - \theta_1 e_{t-1} + e_t \,. \quad (84)$$

The model consists of one previous value of the forecast y_{t-1}, and a previous value of the error term e_{t-1}

Using the notation B, eqn. 84 can be written as

$$(1 - \phi_1 B)\, y_t = c + (1 - \theta_1 B) e_t \quad (85)$$

For higher orders, the model equation is

$$(1-\phi_1 B-\phi_2 B^2 -.....-\phi_p B^p)\, y_t=c+(1-\theta_1 B-\theta_2 B^2 -......-\theta_q B^q)e_t \quad (86)$$

All these models have parameter restrictions and apply the same as discussed in AR and MA cases (Section 3.1.3.3.1, 3.1.3.3.2).

3.1.3.3.4. Autoregressive Integrated Moving Average (ARIMA) Model

ARMA models can be extended to non-stationary series by applying differencing to the series. The only difference from ARMA model is that the difference is also included in this model. Thus the model equation is

$$(y_t-\phi_1 y_{t-1})(y_t-y_{t-1})=c+ e_t -\theta_1 e_{t-1} \quad (87)$$

Expressing eqn. (87) in terms of B, the equation becomes

$$(1-\phi_1 B)\,(1-B)\, y_t=c+(1-\theta_1 B)\, e_t \quad (88)$$

3.1.3.3.5. Seasonal ARIMA Model (SARIMA)

The only difference of SARIMA as compared to ARIMA is that the seasonality is also included in SARIMA. All the different combinations possible with ARIMA is also possible with SARIMA, including the additional term for seasonality. Thus ARIMA(p,d,q) is expressed as ARIMA(p,d,q)$(P,D,Q)_s$. The additional $(P,D,Q)_s$ takes care for the seasonality and s is the number of periods per season. The orders of non-seasonal auto regression and moving average parameters are given by p and q respectively; P and Q are the respective seasonal auto regression and seasonal moving average parameters. The d denotes non- seasonal and D stands for seasonal differences.

A seasonal MA model with show spike at lag 12 in the ACF and PACF showing an exponential decay is expressed as ARIMA(0,0,0)$(0,0,1)_{12}$. Similarly for AR model ARIMA(0,0,0)$(1,0,0)_{12}$, the expected ACF and PACF will be reverse as that for MA model. Thus the model form is

$$(y_t-\phi_1 y_{t-1})\,(y_t-\Phi_1 y_{t-4})\,(y_t-y_{t-1})\,(y_t-y_{t-4})=(e_t -\theta_1 e_{t-1})\,(e_t -\Theta_1 e_{t-4}) \quad (89)$$

The additional terms $(y_t-\Phi_1 y_{t-4})$, (y_t-y_{t-4}) and $(e_t -\Theta_1 e_{t-4})$ are due to seasonality and the additional terms correspond to seasonal AR model, seasonal difference and seasonal MA model respectively.

Expressing eqn. (89) in terms of B, the equation becomes

$$(1-\phi_1 B)\,(1-\Phi_1 B^4)\,(1-B)\,(1-B^4)\,y_t = (1-\theta_1 B)\,(1-\Theta_1 B^4)\,e_t \qquad (90)$$

where Φ_1 and ϕ_1 are the seasonal and non-seasonal autoregressive parameters, respectively. Θ_1 and θ_1 are the seasonal and non-seasonal moving average parameters, respectively.

Most of the steps used in the development of ARIMA models are discussed so far. Using different combinations, many number of models can be framed and one needs to find out an appropriate model from the large number. This is attained by the following steps.

3.1.3.4. Identification of the Model

The first requirement is that the data series should be stationary. This can be checked by plotting ACF and PACF. Sometimes an unusual observation may be present and in such cases transformation is required to stabilize the data. A stationarity in variance is obtained by logarithmic or power transformation of the data. The stationarity can also be obtained by differencing the series, if the data has seasonality present in it. This will take care of any non-stationarity due to mean. The presence of seasonality will give autocorrelations at seasonal lags. Sometime a MA or AR model is revealed from ACF and PACF plots. A data with no significant autocorrelation after lag q suggests a MA (q) model and data with no significant partial autocorrelation after lag p suggests an AR (p) model. If both AR and MA models are not clear then mixture of these models are considered. At some occasions, the seasonal aspects of MA and AR models also required to be taken care of, which makes the task of model identification difficult.

3.1.3.5. Estimation of the Parameters

The coefficients (parameters) of the selected model need to be estimated. For regression model a least square method is suitable. The same is applicable for ARIMA models. But for models with MA part, an iterative method is used. A preliminary estimate is made and it is refined through iteration until the sum of the squared errors is minimum. Another frequently used method is the maximum likelihood estimation. The likelihood (L) of the given data is proportional to the probability of obtaining the data from the given model. In this method, the values of the parameters are found iteratively, in such a manner that the estimated values maximize the likelihood. The estimated parameters are also checked for its Z value (ratio of estimate to its standard error) and P value. If P value is large (insignificant) then the model can be improved by dropping such terms.

Sometimes there is a chance for the identification of more than one model and it will be difficult to decide on the best model. An important criterion to choose the best model from the group of models is to find the sum of squared errors of all models. A model with least value for the squared errors or largest value of likelihood is the best model. But when the number of terms in the model is increased, the error becomes less and the likelihood becomes greater. In order to solve such problem, the additional term in the model is penalized. If no improvement is found in the likelihood upon adding a term then such a term is dropped from the model. There are different procedures used for penalize the likelihood. One common method is Akaike's Information Criterion (AIC) proposed by Akaike (1974). The AIC is a combination of two conflicting factors: (1) the mean square error, and (2) the number of estimated parameters of a model.

$$AIC=-2\log L+2m \tag{91}$$

where L the likelihood and m the number of terms estimated in the model.

$$AIC\approx n(1+\log(2\pi))+n\log\sigma^2+2m \tag{92}$$

where n is the number of observations in the series and σ^2 is the variance of the residuals.

Another procedure utilized is the Schwarz Bayesian Information Criterion (SBIC), which is based on likelihood function and it is a function of the parameters of a statistical model. Generally, a model with smallest value of AIC and SBC is considered as the best model.

3.1.3.6. Diagnostic Checking of the Model

In this method the residuals of the model is checked to find out whether any pattern is present in it. The analysis of ACF and PACF plots helps to understand the presence of patterns in the residuals. No significant autocorrelations or partial correlations will be present for the residuals of a good model. Also, the residuals should be white noise. The number of residuals from the model will be different from the number of observations because of differencing process. The number of residuals is given by n-d-sD, where n is the number of observations, d and D are the degrees of non-seasonal and seasonal differencing and s is the number of observation per season. In addition to the ACF and PACF, the standardized residuals are checked for outliers. Also, the Portmanteau test is applied to the residuals and an insignificant value indicates a white noise series.

3.1.3.7. Forecasting from the Model

Once a model has been finalized. It is used to give forecasts i.e. y_{t+1}, y_{t+2} etc

Consider a simple equation of the form

$$(1\text{-}B)y_t=(1\text{-}\theta_1 B)e_t \tag{93}$$

Expanding eqn. (93) gives

$$y_t\text{-}y_{t-1}=e_t\text{-}\theta_1 e_{t-1} \tag{94}$$

For forecasts the time should be incremented. Thus, eqn. (93) becomes

$$y_{t+1}\text{-}y_t=e_{t+1}\text{-}\theta_1 e_t \tag{95}$$

In eqn. (95), the value at e_{t+1} is not known. It is therefore assumed to be with zero and the value of y_{t+1} is calculated. This process is then repeated for y_{t+2}, y_{t+3} etc

The various steps involved in a model development are now described with an example MOM9. The ARIMA model is developed to forecast the rainfall over Jordan (Momani 2009). The data used for the study is a monthly rainfall series over the Amman Airport Station. Figures 8 and 9 show ACF and PACF of the original non-stationary and stationary data series, respectively. From Figure 9, one autoregressive and one moving average parameters and the exponential decay of ACF starting at lag 12 can be noticed. The PACF also have exponential decay starting at lag 12. Thus an ARIMA(1,0,1) $(1,1,1)^{12}$ model with one seasonal and one non seasonal data can be identified. In order to make sure that the selected model is good, the residuals are plotted. Figure 10 shows the ACF and PACF of residuals from the model. As no patterns are observed, the parameters are now estimated for this selected model. The parameters for the model are estimated in such a way to reduce the sum of squared errors.

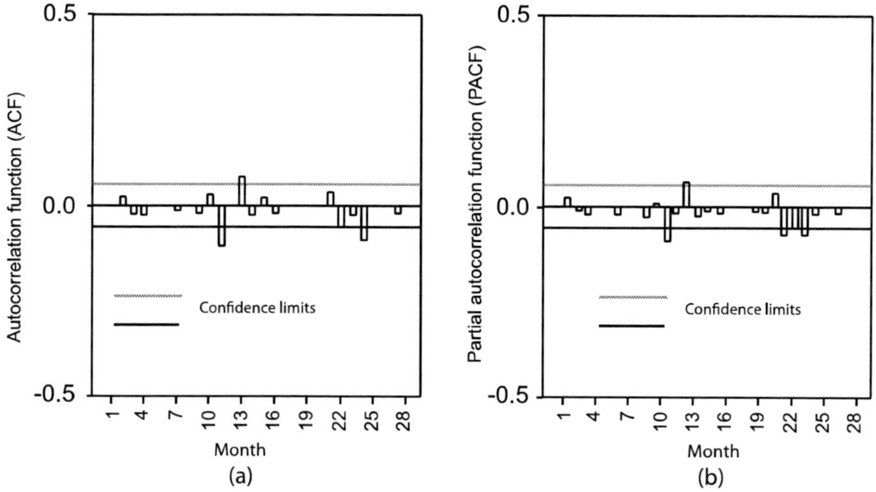

Figure 10. Plots of (a) Autocorrelation and (b) Partial Correlation Function for the residuals of stationary data. (Adapted from Momani 2009).

The basic model form is

$$(1-\phi_1 B)\,(1-\alpha_1 B^{12})\,(1-B^{12})y_t = (1-\theta_1 B)\,(1-\gamma_1 B^{12})e_t \qquad (96)$$

The sum of the squared errors for this model, ARIMA $(1,0,1)\,(1,1,1)^{12}$ is given to be 655288 and the estimated parameter values are given in Table 5.

From Table 5, it is clear that the t value corresponding to γ_1 is very high and all other parameters have low t values. The data series has 936 observations and reducing the number of parameters corresponding to p, q, P and Q resulted in 932 observations. The t value for 932 observations at 5% level significance is determined, which is $t_{0.05}(932)=1.645$. The value obtained is compared with the t value (from statistical tables) for the parameters.

Table 5. Estimated model parameters. (Adapted from Momani 2009)

Parameter	Parameter value	t-value
ϕ_1	0.4308	0.40
α_1	-0.0267	-0.81
θ_1	0.4077	0.37
γ_1	0.9805	200.31

Table 6. Parameter correlation matrix of the selected ARIMA model. (Adapted from Momani 2009)

Parameter	ϕ_1	α_1	θ_1	γ_1
ϕ_1	1.000	-0.058	1.000	-0.008
α_1	-0.058	1.000	-0.060	0.057
θ_1	1.000	-0.060	1.000	-0.008
γ_1	-0.008	0.057	-0.008	1.000

Table 7. Parameter value and t value for the ARIMA model. (Adapted from Momani 2009)

Parameter	ARIMA $(1,0,0)(1,1,1)^{12}$		ARIMA $(0,0,1)(1,1,1)^{12}$	
	Parameter value	t-value	Parameter value	t-value
ϕ_1	0.0714	2.15		
α_1	0.0025	0.07	0.0021	-0.06
θ_1			0.0037	-1.92
γ_1	0.8909	55.33	0.8902	55.04
Sum of squared residuals	737590	737984		

Now the relevance of each of the terms in the model is checked by its correlation. The parameters $\phi 1$ and θ_1 have high correlation (Table 6). So, one of the parameters could be eliminated. The sum of squares for both the cases ARIMA$(1,0,0)(1,1,1)^{12}$ and ARIMA$(0,0,1)(1,1,1)^{12}$ are determined. The sum of squared errors corresponding to these two models is 737590 and 737984, respectively. Table 7 shows the parameter value and t value for both the models. Both the model exhibits similar characteristics and the ARIMA$(1,0,0)(1,1,1)^{12}$ are chosen for further analysis. Now the correlation matrix is analysed to check whether the model can be further improved by dropping predictors that have high correlation. Table 8 shows the correlation matrix for the data considered and no significant correlation is observed.

Table 8. The parameter correlation matrix for the ARIMA $(1,0,0)(1,1,1)^{12}$ model. (Adapted from Momani 2009)

Parameter	ϕ_1	α_1	γ_1
ϕ_1	1.000	0.100	0.105
α_1	0.100	1.000	0.442
γ_1	0.105	0.442	1.000

A good model will have least number of predictors and least squared errors. Therefore, this possibility is checked by dropping the P parameter from the model and the least squared value is determined. The value obtained is 737593 which is not significantly different from ARIMA$(1,0,1)(1,1,1)^{12}$. The t-value obtained for the model parameters are given in Table 9 and the values are higher than $t_{0.05}$ (934)=1.645. Now the ACF and PACF of residuals is plotted to confirm that the selected model is good. Figure 11 shows no pattern in the ACF and PACF plots. Using the Chi-square statistics (χ^2) the goodness of fit was tested. The value of χ^2 for the autocorrelation upto lag 24 is 26.9 and the value obtained 22 (24-p-q) degrees of freedom is 33.92. The value obtained for the model is greater than for the observed value of 26.9. Therefore the residuals are white noise and the selected model is meaningful.

Table 9. Parameter correlation matrix for the ARIMA $(1,0,0)(1,1,1)^{12}$ model. (Adapted from Momani 2009)

Parameter	Parameter value	t-value
ϕ_1	0.15	4.4
γ_1	0.88	44.8

The model form for the ARIMA $(1,0,0)(0,1,1)^{12}$ can be written as

$$(1-\phi_1 B)(1-B)^{12} y_t=(1-\gamma_1 B^{12})e_t \tag{97}$$

Expanding in terms of y, eqn. (97) becomes

$$y_t=(1+\phi_1)y_{t-12}+\phi_1 y_{t-1}+e_t-\gamma_1 e_{t-12} \tag{98}$$

y_t can be obatined by substituting the estimated values.

Similarly, the forecast is obtained by substituting the estimated values and finding the value for y_{t+1}. As the error e_{t+1} is not known, it is usually given the value zero.

Table 10. Ten parameters identified for framing the multiple regression model

No.	Parameter	Month	r_{xy} (1983-2002)
P1	Arabian Sea SST	Jan., Feb.	0.55
P2	Eurasian Snow cover	Dec.	-0.46
P3	NW Europe Temperature	Jan.	0.46
P4	Nino3 SST anomaly (Previous year)	Jul.-Sep.	0.42
P5	South Indian Ocean SST Index	Mar.	0.47
P6	East Asia Pressure	Feb., Mar.	0.61
P7	Northern Hemisphere 50 hPa wind pattern	Jan., Feb	-0.51
P8	Europe pressure gradient	Jan.	0.42
P9	South Indian Ocean 850 hPa zonal wind	Jun.	-0.45
P10	Nino 3.4 SST tendency	Apr.-Jun., Jan.-Mar.	-0.46

Adapted from Rajeevan et al 2004.

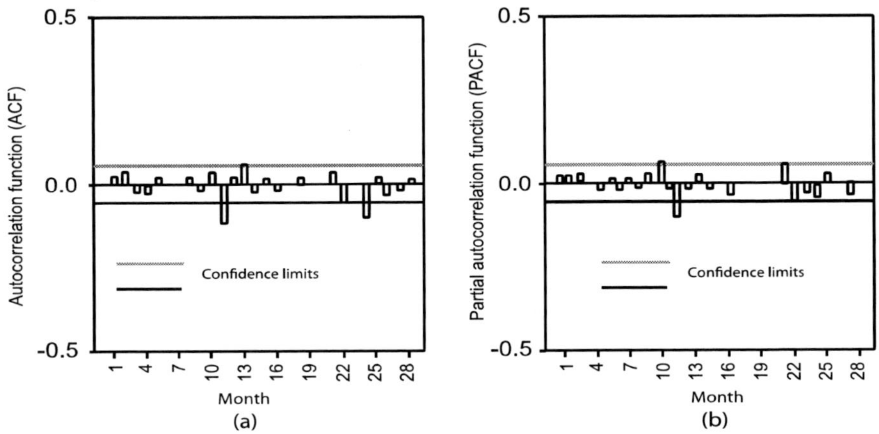

Figure 11. Plots of (a) Autocorrelation and (b) Partial Correlation Function for the residuals of ARIMA (1,0,0) $(0,1,1)^{12}$ model. (Adapted from Momani 2009).

3.2. Non-linear Methods

Nonlinear regression is a form of regression analysis in which observational data are modeled by a function which is a nonlinear combination of the model parameters and depends on one or more independent variables. The data are fitted by a method of successive approximations. Different nonlinear functions are exponential include exponential functions, logarithmic functions, trigonometric functions, power functions, Gaussian function, and Lorenz curves. Functions, such as the exponential or logarithmic functions, can be transformed to linear form and linear regressions are performed but with care.

In general, unlike the linear regression, there is no closed-form expression for the best-fitting parameters. Usually numerical optimization algorithms are applied to determine the best-fitting parameters. Again in contrast to linear regression, there may be many local minima of the function to be optimized and even the global minimum may produce a biased estimate. In practice, estimated values of the parameters are used, in conjunction with the optimization algorithm, to find the global minimum of a sum of squares.

Numerous non-linear models are available. In this section, only the power regression, projection pursuit regression and neural networks are discussed.

3.2.1. Multiple Power Regression

The power regression model includes the non-linear interaction of independent variables with the dependent variable. The power regression model has the following mathematical form

$$R = C_0 + \sum_{1=1}^{n} C_i X_i^{P_i} \qquad (99)$$

where R is the dependent variable, X the independent parameters, C_i and P_i denote the model constants. The values for C_i and P_i are determined by iteration along with the least square fitting method. The iterations are carried out at very small increments. The best model is developed by including suitable parameters in the decreasing order of their correlation coefficient. The parameters are also checked for the consistency in the relationship by using sliding correlation as discussed in previous Section 3.1.2.2 of the chapter. The parameters are standardized before including into the model; otherwise certain scaling constants need to be used. The value of the power is a measure of the nonlinearity, greater the value, greater the nonlinearity.

In the example (RAJ4) considered (Rajeevan et al 2004), the model utilized ten standardised parameters (refer table 10 for parameters used) where the P value varies between 2 and –2. The sliding correlations for the predictors were considered for choosing the ten predictors (Figure 12). The model claims to account for the non-linear interactions of different climatic forcing with the rainfall of the region to be forecasted. The model is developed for the period 1958-95 and tested during 1996-2002. Tables 11(a) and 11(b) gives the forecast and the error of forecast for the ten parameter model. The RMSE during the period 1996–2002 for the 10-parameter model is 6% and the model error is 4%. The model was used for operational forecasting in India.

3.2.2. Projection Pursuit Regression (PPR)

The projection pursuit (PP) was first proposed by Friedman and Tukey (1974). The basis of the PP technique is the linear projection of the given data. It attempts to identify the nonlinear structures within the projections. The name 'projection pursuit' is associated to the fact that the technique automatically pursues the most 'interesting' projections. Huber (1985) proposed PP as an alternative to the traditional principal component technique to reduce the dimensionality in a data set into a lower-dimensional one. One such application of this method was the Projection Pursuit Regression (PPR). This new method of nonparametric MLR was proposed by Friedman and Stuetzle (1981). Chan et al. (1998) demonstrated that PPR is superior to traditional MLR analysis, because the PPR is capable of identifying nonlinear relationships between the predictand and the predictors. Chan and Shi (1999) used this technique to develop models for forecasting of summer monsoon rainfall over south China successfully. The PPR is based on the iterative estimation of linear combinations of the original predictor variables and the corresponding smooth functions that describe the relation between the projection and the response.

Table 11(a). Forecast and error for the 10-parameter model for the period 1958-95. (Adapted from Rajeevan et al 2004)

Year	Actual %	Forecast %	Error %
1958	110	112	-2
1959	114	117	-3
1960	101	103	-2
1961	122	114	8
1962	97	102	-5
1963	98	97	1
1964	110	116	-6
1965	82	84	-2
1966	87	89	-2
1967	100	102	-2
Year	Actual %	Forecast %	Error %
1968	90	86	4
1969	100	102	-2
1970	112	107	5

Year	Actual %	Forecast %	Error %
1971	104	102	2
1972	76	76	0
1973	108	104	4
1974	88	85	3
1975	115	106	9
1976	103	94	9
1977	104	105	-1
1978	109	102	7
1979	81	85	-4
1980	104	104	0
1981	100	102	-2
1982	86	88	-2
1983	113	108	5
1984	96	102	-6
1985	93	95	-2
1986	87	90	-3
1987	81	79	2
1988	119	117	2
1989	101	104	-3
1990	106	107	-1
1991	91	91	0
1992	93	104	-11
1993	100	105	-5
1994	110	100	10
1995	100	105	-5

Table 11(b). Forecast error for the 10-parameter model for the period 1996-2002. (Adapted from Rajeevan et al 2004)

Year	Error (%)
1996	+3
1997	+1
1998	-3
1999	-7
2000	0
2001	0
2002	-15

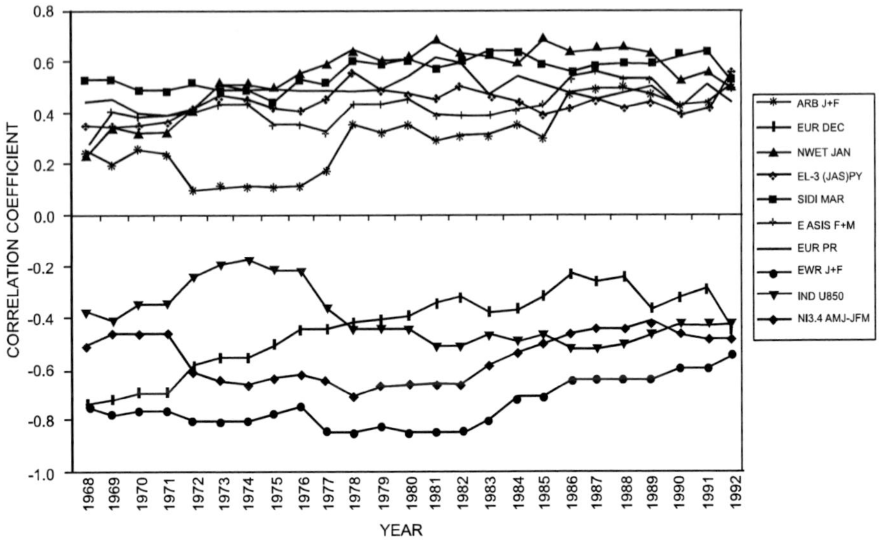

Figure 12. Twenty-one-year moving correlations of the ten parameters with southwest monsoon rainfall (June–Sept). The year indicates the central year of the 21-year period. (Adapted from Rajeevan et al 2004)

The PPR has the form

$$Y(z_1, z_2, z_3\ldots., z_p) = \sum_{m=1}^{M} f_m\left(\alpha_m^T Z\right) \qquad (100)$$

where Z is the (n×p) matrix of 'p' predictor variables and α_m^T is the (p×1) row vector. The projection $\alpha_m^T Z$ gives the linear combination of the predictor variables. The smooth f_m applied on this projection are the single-valued (ridge) functions of a single variable. M is the total number of projections and smooths needed to be able to describe variation in a response variable. The vector α_m^T is fitted by nonlinear fitting such as Gauss–Newton and the function f_m is fitted by the curve-fitting such as splines. Friedman's initial implementation of PPR used super-smoother (Friedman 1984) for fitting f_m. Roosen and Hastie (1994) used smoothing splines for the same purpose. The PPR is identical to a three-layer perceptron or feed forward neural network (Bishop 1995); details given in the Section 3.2.3.1. In neural network, the function f_m is called activation functions. The only difference is that in neural networks the function f_m is the same for all the nodes, while in PPR it can differ for each of the M projections; its format is more flexible and data-driven.

An iterative algorithm defines the exact fitting of the model. The fitting procedure alternates between an estimation of the vector α_m^T and an estimation of function f_m. The function f_m is continuous in the model, but in the algorithm only its values at the data points are used. The fraction of unexplained variance that is explained by adding the term $f_m(x)$ can be used as a criterion of fit (figure of merit):

$$I(\alpha_m) = 1 - \sum_{i=1}^{n}(r_i - f_m(\alpha_m^T Z))^2 \Big/ \sum_{i=1}^{n} r_i^2 \tag{101}$$

where r_i is the residual. The M projections and smooths are produced in a stepwise procedure. The iteration process is initiated by setting $r_i = Y_i$ (assuming that the response is centered: $\sum Y_i = 0$). The vector α_m^T and the corresponding smooth $f_m(x)$ that maximize $I(\alpha_m)$ are selected and the process is terminated when $I(\alpha_m)$ is smaller than a user specified threshold. As long as $I(\alpha_m)$ is not smaller than a threshold value, the residuals are calculated as $r_i = -Y_i - f_m(\alpha_m^T Z)$. These residuals are then used as the response values in the next step of the stepwise procedure.

In the example RAJ6, a six parameter PPR model (PPR-I) was developed by Rajeevan et al 2006, to give the first stage forecasts of Indian Summer Monsoon rainfall (ISMR) in April using the selected predictors. The predictors identified are North Atlantic SST anomaly (December & January), Equatorial SE Indian ocean SST anomaly (February & March), East Asia surface pressure anomaly (February & March), Europe land-surface air temperature anomaly (January only), Northwest Europe surface pressure anomaly tendency (December-February (0)-September-November (-1); where (0) and (-1) denotes the current and the previous year respectively) and warm water volume (WWV) anomaly (February & March). In this study, two projections were used and therefore, the predictand was expressed as the sum of the functions of first two projections. The model was trained over the period of 1968–1980.The independent predictions are made by sliding the training period with an optimal window period of 23 years. The linear relationship between the first projection predictors and ISMR is represented by $X_1 - \alpha_1^T$ and that between residual $r(z) = Y - f_1(\alpha_1^T Z)$ and the second projection ($X_2 = \alpha_2^T Z$) represents the nonlinear relationship, removing the contribution of X_1 in the ISMR. This is one of the important advantages of PPR models in comparison to the models based on linear regression. The model is trained for the period of 1968–1980, for forecasting ISMR in 1981. Thus, the first two most interesting projections of the predictor set represented the linear and nonlinear relationships, respectively, between the ISMR and the predictors.

The predictions by PPR-I model was good during the recent years. The RMSE of predictions by the PPR model for the common period (1981-2004) were smaller than that of the predictions by climatology based model. The RMSE of PPR-I for the period of 1981–2004 was 6.75% LPA, respectively. They were smaller than that of the forecasts based on climatology alone.

3.2.3. Artificial Neural Network (ANN)

The artificial neural networks are based on the design to mathematically model the functions of brain such as learning, pattern recognition, classification and generalization. When applied to the time series a non-linear forecasting method is obtained. This method requires a large number of observations such that a complex model can be fitted to the observations. The great advantage of this method is that it is useful for non-linear approximation. A neural network is a network of neuron-like units organized in layers. The bottom layer consists of input units and the top layer consists of output units. The units in each layer is linked to the units in the higher layer.

A neural network is specified using the number of layers and units in the network. The layers of the neurons are called processing elements. The inputs are then combined to give the output and the amplitude of the output is controlled by the activation function. A nonlinear activation function is a combination of a linear function of the inputs followed by a non-linear squashing function. The activation function is basically of three types and it limits the output value to be between 0 and 1 or –1 and 1. First is a Threshold function, which takes a value of 0 if the summed input is less than a certain threshold value, and 1, if the summed input is greater than or equal to the threshold value. Second is a Piecewise-Linear function. This function takes the values of 0 or 1, but can also take on values in between, depending on the amplification factor in a certain region of linear operation. Third function is the sigmoid function. This function can range between 0 and 1, but it also takes the -1 to 1 range. An example of the sigmoid function is the hyperbolic tangent function. The output is measured for forecast accuracy, represented by cost function, which in turn minimized by using suitable training algorithm. Thus as of other models, the input is the same for this method, but a network is used instead of a model and instead of parameters weights are used and the network is trained instead of estimating the parameters. The weights are attached to each input and are selected in such a way that the MSE is minimum. A neural network consists of non-linear hidden layers, which is connected to the input and output linear layers. A disadvantage of this method is that, it does not require explicit model and limits the understanding of the data. So they provide black box approach to forecasting. In this section, a feed forward neural network with error back propagation algorithm is discussed.

3.2.3.1. Feed-Forward Neural Network (FFNN)

The feed-forward neural network is the simplest of artificial neural network. In this method the information moves only in one direction from input to output through hidden layer. There are no cycles in the feed forward network, whereas recurrent networks are identified with the presence of cycles. Thus for feed forward network, each neuron in a layer passes its output to every neuron in the next layer hence the name feed forward neural network. The network output is obtained from the output of the output layer. A three layer FFNN is shown in Figure 13. The FFNN has the ability to approximate any non-linear relationship between the input and the output. This method is useful in satellite data retrieval, interpretation of satellite imagery, process identification and control, fault detection and diagnosis etc. The neural network is made to learn the relationship between the input and the output by means of network training and the commonly used training algorithm with FFNN is the error-back-propagation (EBP) (Rumelhart et al 1986 a, b).

The EBP needs the activation function used in the neural network to be differentiable. There are two important steps in EBP; one is the propagation and other the updation of weights. In the propagation step, a forward pass using an input pattern propagates through the network and produces an actual output based on prediction error and a backward pass uses the desired output and propagates backward updating the weights. For this, the network output (output from hidden layer) is compared with the desired value and the error is calculated. The calculated error is called the prediction error, which is used to correct the weight between the hidden and output layer and subsequently between the hidden and input layers. Thus, the weights are adjusted until a pre-specified error criterion if fulfilled. At this point the weights are said to be optimised.

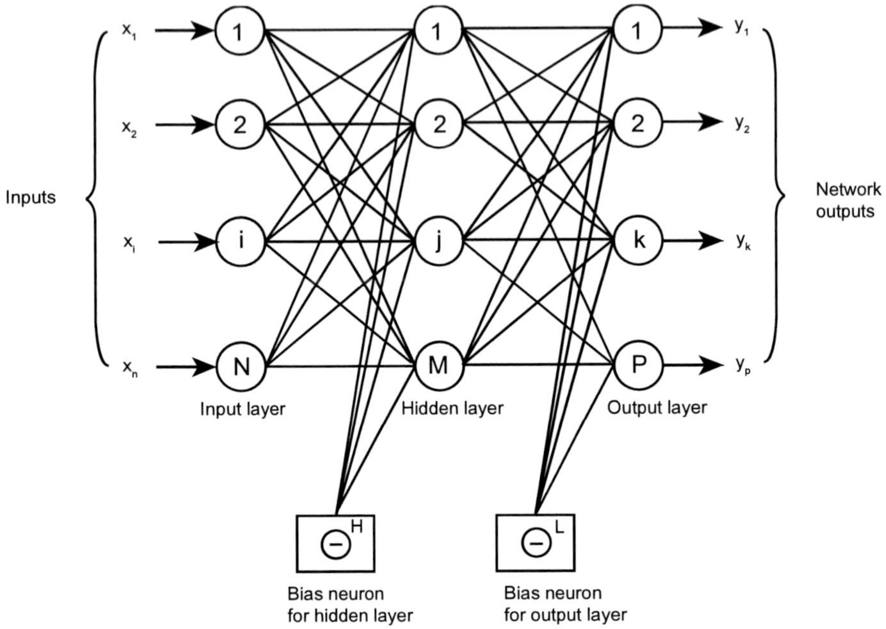

Figure 13. A three layer feedforward neural network. (Adapted from Venkatesan et al 1997).

After training, the EBP network can predict output for the given input. But the network should be able to give output for a new set of input values which are not included in the training set. This process of accurately predicting the output for a new set of input vectors is called generalization. As for other models the network is also trained (learning) for a period and tested for another period. The method sometimes has slow convergence and entrapment into a local minimum. The training is made faster by including the momentum term and avoids local minimum. The demerit of this method is overfitting, which results from the training over a large number of training iterations and more number of hidden neurons. When the network is trained for large number of iterations then the network learns all the minute details and when new inputs are given, it fails to generalize in a better manner and gives erroneous results. Thus the problem of overfitting is handled by stopping the training set when RMSE with respect to the test set is minimum.

A three layer FFNN using the EBP algorithm is explained with the example, VEN97 (Venkatesan et al 1997). To start with the network some random values are assigned as the weights and the input predictors (vectors) are given to the input layer. When all inputs are weighted and are added up, the input to the hidden layer is ready. This can be represented as

$$\mathrm{Net}_{qj}^{H} = \sum_{i=1}^{N} W_{ji}^{H} x_{qi} + \Theta_{j}^{H}, j = 1, M \tag{102}$$

where Net_{qj}^{H} is the weighted sum for the jth hidden layer neuron when q[th] input pattern has been applied; W_{ji}^{H} is the weight between i[th] input layer and j[th] output layer neuron; x_{qi} is the element of input pattern X_q; Θ_{j}^{H} is the bias weight for j[th] hidden layer neuron; N refers to the number of input layer neuron. Here the input for the model consists of two predictors: (1) ridge 500mb, and (2) Bombay pressure tendency (DSLP). The model inputs are chosen for the period 1939-1994. Thus a total of 56 patterns are present which is divided into 49 (training) and 7 (test) patterns respectively. The observations for the corresponding years are given in Table 12 and patterns considered for the test are marked.

Using an activation function (sigmoid activation function), the input to the hidden layer is transformed in order to get the output from hidden layer neuron.

$$I_{qj}^{H} = \frac{1}{1 + \exp(-\mathrm{Net}_{qj}^{H})}; j = 1, M \tag{103}$$

where I_{qj}^{H} is the output from jth hidden layer neuron when q[th] pattern is presented. Again the weighted sum of inputs for the output layer is determined (as for the hidden layer) and transformed using an activation function. The input and the outputs to the network were normalized by applying linear transformations. The network training consisted of 15,000 iterations and a single iteration involved the forward and backward passes (equal to the number of input-output patterns in the training set).

Table 12. The predictor values and the corresponding rainfall for the period 1939-1994

Year	Ridge	DSLP	All India Rainfall (mm)	Year	Ridge	DSLP	All India Rainfall (mm)
1939	14.0	4.4	789.7	1967	17.5	5.6	860.1
1940	15.3	5.0	853.5	1968*	12.5	5.7	754.6
1941	11.2	2.8	728.4	1969	17.3	5.2	831.0
1942	17.5	1.8	957.9	1970	15.8	1.8	939.8
1943*	16.0	2.5	868.4	1971	16.8	2.2	886.8
1944	14.5	3.0	920.6	1972	11.0	4.8	652.9
1945	16.7	3.8	911.1	1973	16.8	1.7	913.4
1946	17.3	4.1	903.8	1974	13.5	4.8	748.1
1947	18.0	2.1	945.6	1975	17.5	1.3	962.9
1948	14.5	1.7	874.0	1976*	17.0	4.8	856.8
1949	17.0	2.8	904.1	1977	14.0	3.4	883.2
1950	17.0	3.0	877.0	1978	14.0	1.9	909.3
1951	12.0	4.1	739.1	1979	12.5	3.7	707.8
1952*	13.5	3.6	793.1	1980	15.0	3.8	882.8
1953	17.0	2.0	923.2	1981	17.0	4.8	852.2
1954	16.5	2.8	885.6	1982	11.3	3.6	735.4
1955	15.5	0.6	930.4	1983	14.5	0.3	955.7
1956	17.5	3.5	983.3	1984*	14.8	3.0	837.7
1957	16.0	3.0	788.7	1985	14.7	0.2	759.8
1958	17.0	0.4	889.4	1986	15.0	4.6	743.0
1959	16.0	2.8	944.1	1987	14.0	4.4	697.3
1960*	16.7	2.0	839.8	1988	14.5	2.3	961.5
1961	15.0	1.9	1020.3	1989	16.5	3.1	866.7
1962	14.8	4.7	809.8	1990*	15.0	0.1	908.7
1963	13.5	2.4	857.9	1991	14.5	3.6	785.2
1964	18.3	1.9	922.6	1992	13.5	0.1	785.5
1965	14.0	5.0	709.4	1993	14.5	5.6	894.0
1966	13.5	2.5	739.9	1994	14.0	4.1	938.3

* Represents the data of test set (Adapted from Venkatesan et al 1997).

The error associated with each output layer neuron is given as

$$\delta_{qk}^{L} = (Y_{qk} - I_{qk}^{L})[I_{qk}^{L}(1 - I_{qk}^{L})]; k = 1, P \qquad (104)$$

where δ_{qj}^{L} is the error term Y_{qk} is the desired output of neuron k for pattern q and I_{qk}^{L} is the actual output of the kth output layer neuron. Similarly the error

δ_{qj}^{H} for the hidden layer is also calculated and the weights are updated between the output and the hidden layer.

$$\delta_{qj}^{H} = \sum_{k1=}^{P} \delta_{qj}^{L} W_{qj}^{L} [I_{qj}^{H}(1 - I_{qj}^{H})]; j = 1, M \tag{105}$$

$$W_{kj}^{L}(t+1) = W_{kj}^{L}(t) + \eta\delta_{qk}^{L} I_{qj}^{L} + \alpha[W_{kj}^{L}(t) - W_{kj}^{L}(t-1)]; j = 1, M \tag{106}$$

where t denotes the number of iterations, η represents the learning coefficient (0< η<1) and the α of the momemtum term denotes the momentum coefficient (0< α<1)

Now the weights between the hidden and the input layer is updated

$$W_{ji}^{H}(t+1) = W_{ji}^{H}(t) + \eta\delta_{qj}^{H} x_{qi} + \alpha[W_{ji}^{H}(t) - W_{ji}^{H}(t-1)]; i = 1, N \tag{107}$$

Then a new input vector is given and repeated the above steps until the network weights are optimized. In the given example, VEN97 the network uses a momentum term and logistic sigmoid transfer function at the hidden and output layer. After a single iteration the RMSE for the training and test sets were calculated. This procedure with random weights is repeated until a minimum RMSE for the test is obtained. This is calculated as:

$$E_{tst} = \sqrt{\sum_{q=1}^{N_{tst}} \sum_{k=1}^{P} \frac{(y_{qk} - I_{qk}^{L})^2}{N_{tst} P}} \tag{108}$$

where E_{tst} is the error for the test and N_{tst} signifies the number of test set patterns. The trial that give minimum RMSE is called error-surface-exploration. Figure 14 shows the RMSE corresponding to training and test set as a function of number of neurons in the hidden layer. From Figure 14, it is clear that a minimum of RMSE is obtained for the six hidden neurons. Therefore, the optimal weights for the six hidden units are determined as shown in Table 13(a) and 13(b).

Thus, the network configuration for this model is 2:6:1. The RMSE for the training and the test are 66.83mm and 24.53 mm, respectively, and the corresponding correlation coefficient is 0.7 and 0.91.

In Section 3, six quantitative forecast methods have been discussed. In Section 4, qualitative forecasts methods are discussed.

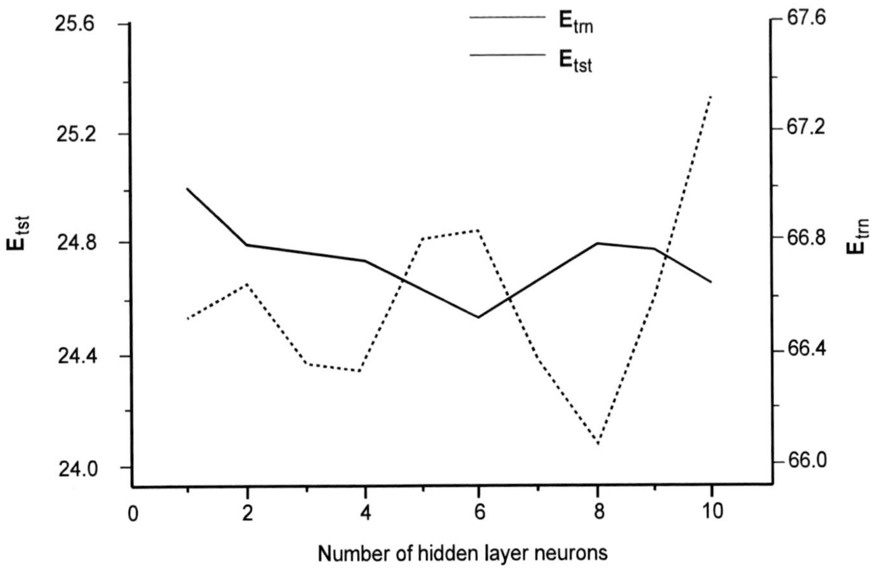

Figure 14. RMS error wih respect to training and Test sets as a function of number of hidden layer neurons. (Adapted from Venkatesan et al 1997).

Table 13a. Optimal weights for the model (Adapted from Venkatesan et al 1997)

From	To j^{th} hidden layer neuron					
	j=1	j=2	j=3	j=4	j=5	j=6
Input neuron1	-0.294163	-0.435573	0.168131	-1.43422	-1.69509	-0.109301
Input neuron2	0.319022	-0.548771	-0.569852	1.16194	1.02023	-0.0324836
Bias neuron, Θ^H	-0.406677	-0.435841	-0.740890	-0.434811	-0.401620	0.379037

Table 13b. Optimal weights for conections between hidden neurons and output neurons (Adapted from Venkatesan et al 1997)

From	To the output neuron
Hidden neuron 1	-0.484905
Hidden neuron 2	0.389519
Hidden neuron 3	0.978241
Hidden neuron 4	-1.21877
Hidden neuron 5	-1.89753
Hidden neuron 6	0.342731
Bias neuron, Θ^L	0.742981

4. Probablistic Forecasts

Probabilistic forecasting is a technique for weather forecasting that relies on different methods to establish an event occurrence/magnitude probability. This differs from deterministic forecasting which gives definite information on the occurrence/magnitude (or not) of a particular event. Both techniques try to predict events but information on the uncertainty of the prediction is only present in the probabilistic forecast. Probabilistic forecasts allow for optimal decision making for many purposes, including air traffic control, ship routing, agriculture, electricity generation and weather-risk.

4.1. Probablistic Forecast from Ensembles

In this method a large number of multiple regression models are developed from which the best model is identified. The need for such a model is that a single model cannot explain the variability of all the rainfall years. Different models developed with same correlation and RMSE values will give different results (i.e. a model that can predict dry years may not be suitable for wet years). In such situations, a probabilistic forecast from an ensemble of many independent forecasts seems to be a practical option.

As in many other methods, a set of predictors are identified by using some conventional methods. The selected predictors are then used in a multiple regression model. As the attempt is to make ensembles, the training years are not a pre-fixed period. Instead, a random selection of the training years is made over a given period. This enables one to make a large number of models with different training period. The coefficients have been found separately for all the models developed and were subjected to bias removal and variance correction. This is given as

$$F_{inf} = (F - \overline{F}) * (O_{sd} / F_{sd}) + \overline{O} \tag{109}$$

where F_{inf} is the variance inflated bias corrected forecast, F and O are forecasted and observed values, $\overline{F}$ and $\overline{O}$ are their means over the training period and F_{sd} and O_{sd} are the standard deviations respectively.

Sahai et al (2008)(SAH8), used a previously developed empirical model to construct a large ensemble of models to deliver useful probabilistic forecast of All India Rainfall (AIR). Only the global sea surface temperature (SST) was made use as a predictor in the empirical model. The model was developed for a period 1921–1998 and was verified over the period 1999-2008. Further, the model development period was divided into two. Thus, 58 years out of the 78 years were selected as the training period and the remaining 20 years was selected as the test period. Thus, 10,000 random training sets were framed using different combinations of the 58 years. The best predictor set for the model was selected using the leave one out cross-validated stepwise regression method. This procedure was repeated for the randomly selected 10,000 training sets. Such an approach helps to remove the non-stationarity problem of the predictor-predictand relationship. Hindcasts from 10,000 models were obtained for the period 1921-1998 using the leave one out cross validation method and the independent forecast delivered using the multiple regression equation developed for the development period. The variance inflated bias corrections are made, as the models developed basically give forecast near normal with small variance. The uncertainty arising from internal variability and from the decadal variability of the predictor-predictand relationship is implicitly included into the model, which is attained by selecting 1,000 models from 10,000 models based on the performance for the past 30 years.

Hence, the 1,000 hindcasts were obtained and these models were used for verification purpose. Relative Operating Characteristic (ROC) curve and Area Under Curve (AUC) are the measures used for forecast verification. The ROC curves are based on contingency tables obtained by binning the forecast probabilitites into 10 equal bins of 0-10%, 10-20%. 90-100%. The ROC curves are shown in Figure 15. The perfect model will have 1.0 and a climatological model will have 0.5 for ROC AUC.

The categories defined for AIR is given in Table 14. The skill of the probabilistic forecast of the model under consideration is given in Table 15. It is clear that the forecast for the drought years is good.

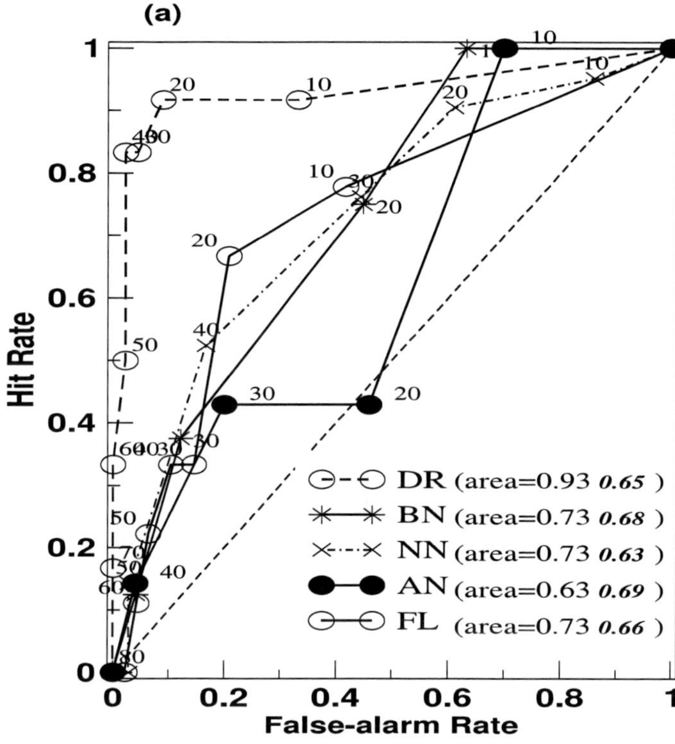

Figure 15. The Relative Operating Characteristic (ROC) curve constructed on the basis of forecast categories described in Table 14 for 1000 hindcast. Area under curve (AUC) for each category is also denoted in braces along with their 95% significance level (in bold italics) for comparison. (Adapted from Sahai et al 2008).

Table 14. Five categories chosen for the Rainfall departure as a Percentage of LTM (Adapted from Sahai et al 2008)

Category	Drought (DR)	Below Normal (BN)	Near Normal (NN)	Above Normal (AN)	Flood (FL)
Limits	≤-10	>-10 and <-4	≥-4 and ≤4	>4 and <10	≥10

Table 15. Probability of Rainfall Occurance in Five Categories. Bold represents the most likelihood forecasted category (Adapted from Sahai et al 2008)

Year	Observed Category	Forecated category				
		DR	BN	NN	AN	FL
1999	BN	32	23	30	13	2
2000	BN	20	24	33	15	8
2001	BN	20	19	38	15	8
2002	DR	53	26	17	3	1
2003	NN	13	8	22	23	34
2004	DR	47	24	25	4	1
2005	NN	9	27	47	12	5
2006	NN	12	48	37	3	0
2007	AN	5	14	49	18	14
2008	-	3	9	38	45	5

For a given year, the RMSE was calculated from the past 30 years and 1,000 models that have lowest RMSE were selected. Then, the average forecast is obtained from the mean for these 1,000 models. This procedure is then repeated to obtain results for all the test years. The model is able to foresee the variability in the observations with sufficient confidence and demonstrate the capability of predicting monsoon droughts with high degree of confidence. The skill scores determined through rxy (0.9) and the RMSE (5%) are high. The probabilistic forecast indicate that for the year 2008, there is a 45% probability for the monsoon to be in the 'Above Normal' category and 38% probability for the 'Near Normal' category, while the ensemble mean obtained from the same model indicate a departure of +3.6% of long term mean which shows that this year will experience a normal to excess AIR.

4.2. Parametric Model

The parametric model is a qualitative method and indicates the likelihood that the monsoon rainfall will be in excess or deficient depending upon the proportion of favourable and unfavourable parameters. Even though the method is qualitative, it was considered to be a powerful tool as the decision can be made in advance if the favourable/unfavourable conditions are higher than a certain percentage. The very basic and important step of this method is the selection of parameter based on the physical linkages of the parameters to the monsoon. The favourable and unfavorable parameters are selected by correlation analysis. The relation that gives a positive correlation is considered favourable and those give negative relations are considered unfavourable and assigned the alphabet F and U, respectively. The percentage of F out of the total number of parameters is found and the percentage of favourable parameters is calculated.

Gowariker et.al 1991, utilized a 16 parameter model which was used for the qualitative analysis of monsoon over India. The 16-parameters used were El Nino (Same Year), El Nino (Previous Year), North India Temperature (March), East Coast of India Temperature (March), Central India Temperature (May), Northern Hemisphere Surafce Pressure anomaly (January to April), 500 hPa ridge (April), 50 hPa Wind Pattern (Winter), 10 hPa Zonal wind Pattern (Jannuary), Northern Hemisphere Temperature (January and February), S.O.I (Spring), Darwin Pressure (Spring), Argentina Pressure (April), Indian Ocean Equatorial Pressure (January to May), Himalayan Snow Cover (January to March), Eurasian Snow Cover (Previous December). The qualitative analysis done for the period 1951-90 was based on the favourable/ unfavourable conditions of these parameters. Whenever 55% or more parameters were favourable, the monsoon was normal.

5. Spatial Field Quantitative Forecasting Techniques

5.1. Principal Component Analysis (PCA)

When a set of p variables is correlated among themselves, the technique of principal component analysis is utilized to reduce the dimensionality of the data. It transforms the original dataset to a new set of uncorrelated variables called principal components (PCs). The new variables are linear combinations of original variables and are derived in decreasing order of importance. The new variables obtained are orthogonal components and account for the maximum variance in the original data. The PCA is also known as empirical orthogonal function (EOF) analysis.

The procedure for EOF analysis can be summarised as follows.

- Take the data matrix X and remove the time mean from the time series.
- Find the covariance matrix $R=X^TX$; (if X is a standardized data matrix then $R=X^TX$ gives a correlation matrix)
- Find the eigen values and eigen vectors of R by solving $RC=C\Lambda$
- Find the biggest eigen values and their corresponding eigen vectors, the EOFs
- Find the expansion coefficients by calculating $\vec{a}_j = X \times EOF_j$

Assuming that the mean from each of the time series in X is removed so that each column has zero mean. Then, the covariance matrix (correlation matrix; for standardized data matrix) of X is formed by calculating

$$R=X^TX. \tag{110}$$

The original data matrix is a (n x p) matrix and the resulting covariance/correlation matrix has an order of (p x p). The eigen value problem is then solved as

$$RC=C\Lambda \tag{111}$$

Λ is a diagonal matrix containing the eigen values λ_i of R. The c_i column vectors of C are the eigenvectors of R corresponding to the eigen values λ_i. Note that the eigen values are the solutions of the determinant equation (Draper and Smith 1981)

$$\left|X^tX - \lambda I\right| = 0 \tag{112}$$

Both Λ and C are of the order (p x p). For each eigen value λ_i, the corresponding eigenvector c_i is determined. Each of these eigenvectors can be regarded as a map. These eigen vectors are the EOFs. The eigenvectors are ordered according to the size of the eigen value. So, EOF1 is associated with the largest eigen value, EOF2 with the second largest eigen value and so on. Each eigen value λ_i gives the fraction of total variance explained by the mode. The eigenvector matrix C has the property that $C'C = CC' = I$. This property implies that the EOFs are uncorrelated over space or the eigenvectors are orthogonal. The principal components (Z) is a (n x p) matrix given as

$$Z=XC \tag{113}$$

5.1.1. Covariance and Correlation Matrix

Usually covariance matrix of the data is used while carrying out an EOF. This can also be done with correlation matrix. In covariance matrix the variables are not standardized whereas in correlation matrix the variables are standardized.

5.1.2. Truncation (Elimination) of Principal Components

The number of PCs that have to be retained/truncated is a matter of debate. The principal components are uncorrelated with one another and can be ordered with respect to the magnitude of their eigen values. The first principal component will account for the largest proportion of variance and hence the eigen value associated with it will be the largest. Some studies are based on the percentage of total variance explained by the principal components. There are different thresholds such as 70% to 90% is found to be reasonable range (Jolliffe 2002). Sometimes a graph called scree plot is used for this purpose. This kind of truncation is based on the shape of the graph of eigen values in the decreasing order as a function of m, which denotes the number of PCs. This is known as eigen value spectrum. It is known as a spectrum as the eigen values represent variance of the PCs. The scree plot is the eigen value spectrum with a linear vertical scale. From the scree plot the point of steeply sloping portion (left) and the shallowly sloping portion (right) are separated and number of PCs that occur to the left of the point is considered. Another method is to consider only those principal components with eigen value greater than one (Draper and Smith 1981). This method is often referred to as the "Kaiser-Gutman Rule" (Loehlin 1998). Some other method selects the first principal components whose combined eigen value product is greater than one (Guiot et al. 1982). As the principal components are orthogonal regressor variables, a criterion to control the order of reduction is the t-statistics given by

$$t=\frac{\hat{\alpha}_j}{S_{\hat{\alpha}_j}}=\frac{\hat{\alpha}_j\sqrt{\lambda_j}}{S} \tag{114}$$

where $S_{\hat{\alpha}_j}$ is the standard error (s.e.) of $\hat{\alpha}_j$ (Myers 1986). In this procedure, t-values should be rank ordered and components should be considered for elimination beginning with the smallest t-value.

5.1.3. Centering and Scaling

When x_{ji} the ith year of observation and j[th] climatic variable in its original units is transformed to x^*_{ji} as

$$x^*_{ji} = \frac{x_{ji} - \bar{x}_j}{S_j} \tag{115}$$

where x represent a set of independent variables

$$S_j = \sqrt{\sum_{i=1}^{n} (x_{ji} - \bar{x}_j)^2} \tag{116}$$

Such a dataset is called centered and scaled.

5.2. Principal Component Regression (PCR)

The Principal Component-Regression (PCR) technique is recommended when there is significant intercorrelation among the predictors. This model is proposed by Draper and Smith 1981. The PCA helps to reduce the degrees of freedom by restricting the number of predictors and thereby avoid intercorrelations between the predictors. To develop a PCR model, the first step is to apply a Principle Component Analysis (PCA) on the predictor set and a few PCs of the correlation matrix that have large variance are selected. The selected PCs are used in a multiple regression model. This method is known as principal component regression.

The general equation for a multiple linear regression model is:

$$y_i = b_0 + b_1 x_1 + b_2 x_2 + \cdots + b_n x_n + e \tag{117}$$

For a principal component regression the x's are replaced by the selected principal components Z, which is a (n x p) matrix. Therefore, eqn. (117) takes the form

$$y_i = b_0 + \alpha_1 Z_1 + \alpha_2 Z_2 + \cdots + \alpha_n Z_n + e \qquad (118)$$

where Z=XC and α=C′b, Z=z_1, z_2,…..,z_p and α=α_1, α_2,…..,α_p is a px1 vector of new coefficients. The insignificant principal components can also be eliminated by applying a stepwise regression. The steps that are followed hereafter is similar to that of multiple regression discussed in Section 3.1.2.

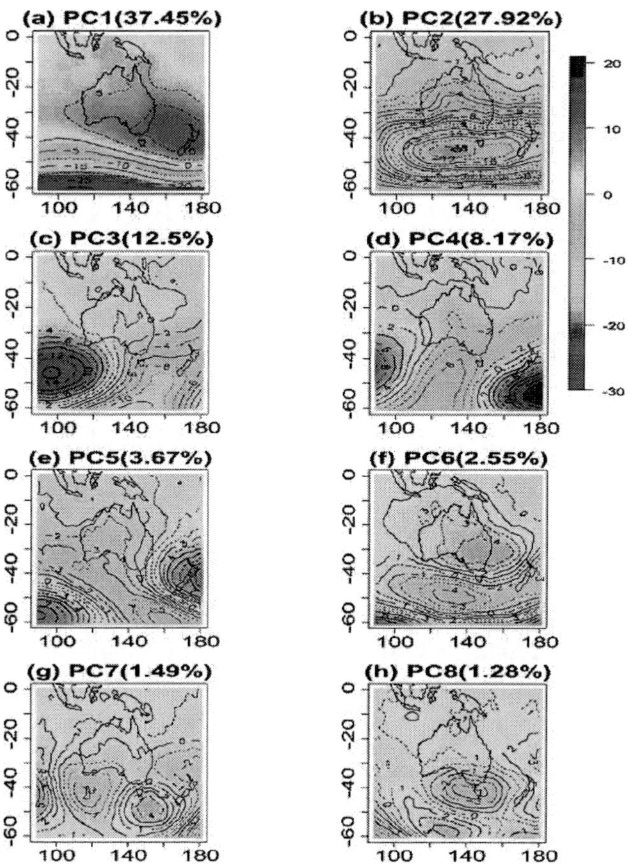

Figure 16 (a-h). The first eight leading spatial patterns of winter MSLP over the study region. Variance explained is given in parenthesis. (Adapted from Li and Smith 2008).

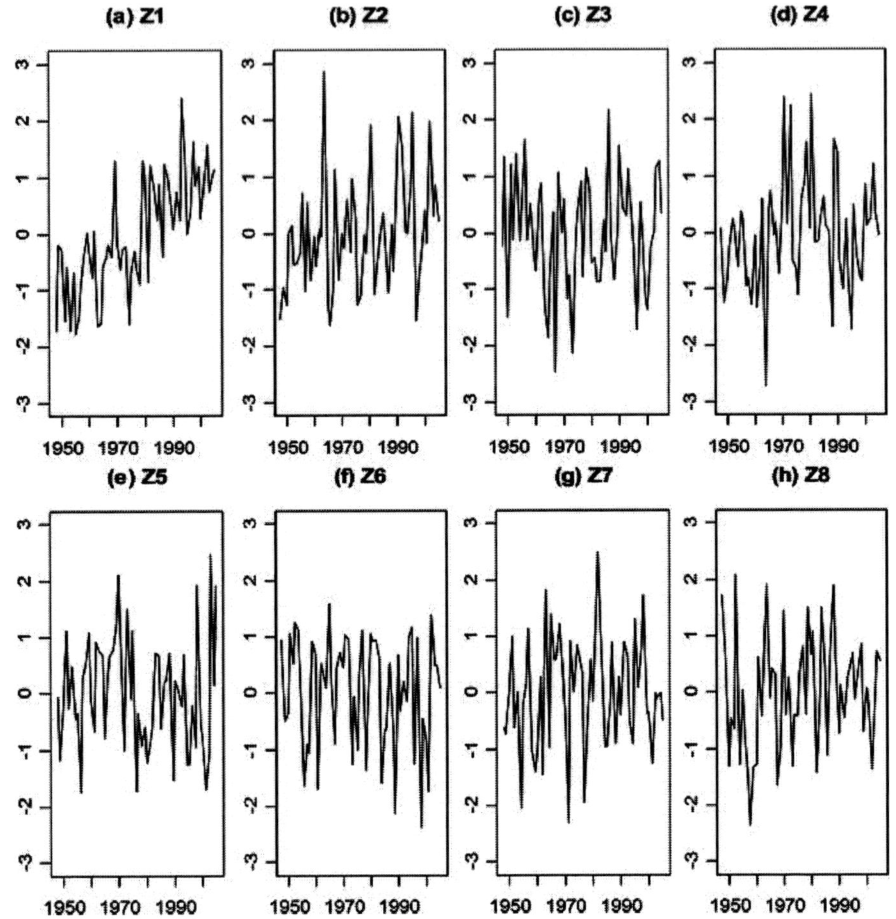

Figure 17 (a-h). The first eight leading temporal patterns of winter MSLP over the study region. (Adapted from Li and Smith 2008).

Table 16. The correlation between leading eight MSLP modes with four target regions identified in the study. Values significant at $p < 0.05$ are in bold (Adapted from Li and Smith 2008)

Region	SOI	SAM	Z_1	Z_2	Z_3	Z_4	Z_5	Z_6	Z_7	Z_8
SWWA	**0.34**	**-0.37**	**-0.40**	**0.37**	**-0.34**	**-0.28**	0.25	-0.17	0.24	0.11
SA	**0.36**	-0.03	-0.25	**0.47**	0.16	**0.44**	0.08	**-0.28**	0.19	-0.03
VIC	**0.43**	-0.19	**-0.36**	**0.59**	0.20	0.18	0.15	-0.18	0.02	0.11
TAS	**0.31**	-0.12	-0.15	**0.69**	**0.31**	0.06	0.21	0.07	-0.17	0.12

The example, LI8 (Li and Smith, 2008) uses an EOF analysis and the resulting PCs are used to develop a PCR model. An index is made by taking the average of June-August mean sea level pressure (MSLP) over the region 60°S-0°, 90°E-180°E. The winter rainfall (June-August) data with a resolution of 0.25° x 0.25° for a 58-year period from 1948-2005 is obtained from National Climate Centre (NCC) of the Bureau of Meteorology (BoM). Applying EOF analysis on MSLP data yields eight PCs that explain 95% of the total variance. The spatial and temporal patterns of the EOF on MSLP are shown in Figures 16 and 17, respectively.

The correlation of each of the eight modes with four sub-regional rainfall indices are given in Table 16. The four target region used in the study are (1) southwest Western Australia —the region southwest of the line joining 30^0S, 115^0E and 35^0S, 120^0E; (2) the south Australian region (SA)—the box area bounded by 32–36^0S, 135–140^0E; (3) the Victorian region (VIC)—the box area bounded by 37–39^0S, 143– 147^0E; and (4) the Tasmanian region (TAS)—the box area bounded by 41–43^0S, 145–148^0E. A PCR model is developed for the four target regions identified in the study.

The PCR utilised in this study gives the basic model form as:

$$y_t = \sum_{m=1}^{M^*} \alpha_m Z_m(t) + \varepsilon_t \tag{119}$$

where y is the estimated rainfall, Z selected principal components, α coefficients obtained from least square estimates, ε the error and M^* the number of PCs. In example LI8, M^* is eight. The PCR model is developed using the selected PCs and trained over the 43-year period from 1948-90 and tested over the 15 years from 1991-2005. A cross validation method proposed by Stone (1974) is used to decide on the number of PCs that describe rainfall time series. During the training period, any improvement in the variance explained is checked. It is found that in most cases, first four modes explain a majority of the variance but for Tasmania only three PCs were required. The total percentage variance explained for each region is 52% for SWWA, 55% for SA, 61% for VIC and 56% for TAS. The PCR predicted and observed rainfall is compared, as shown in Figure 18. The model is assessed in terms correlation coefficient and the ratio of RMSE to climatological JJA rainfall. The results are summarized in Table 17.

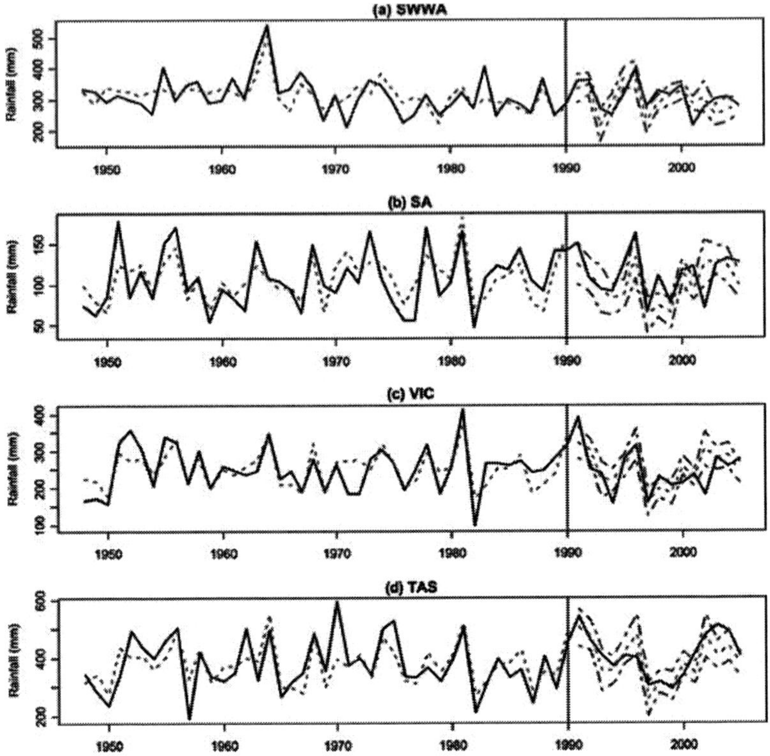

Figure 18 (a-d). Predicted (dash curve) vs. observed JJA rainfall amounts (solid curve) for each of the four regions. The results for 1948–90 correspond to the training period, while the results for 1991–2005 correspond to the verification period. The other dash curves, during the verification period, are upper and lower bands of the 95% confidence interval for the verified predictions. (Adapted from Li and Smith 2008)

Table 17. Correlation between the observed and forecast, ρ (%) is the ratio of RMSE (mm) to Climatology of winter rainfall for the training (1948-90) and verification period (1991-2005) (Adapted from Li and Smith 2008)

Regions	Training period (1948-1990)		Verification period (1991-2005)	
	R	ρ (%)	R	ρ (%)
SWWA	0.72	12	0.69	13
SA	0.74	22	0.61	22
VIC	0.78	15	0.64	19
TAS	0.75	16	0.80	12

5.3. Canonical Correlation Analysis (CCA)

Canonical Correlation Analysis was introduced by Hotelling (1935) and was extended to atmospheric science by Glahn (1968). A CCA deals with the linear relationship of two highly correlated canonical variables, one corresponding to predictor (independent variable) and other the predictand (dependent variable). It is similar to multiple regression except that CCA is multivariate analysis on the independent and dependent variables. The spatial patterns obtained on this analysis, therefore includes a combination of both the fields in such a way that the linear correlation between the two variables is maximized. This linear combination is extracted and the steps are repeated for the residual data, with the constraint that the second linear combination of variables must not correlate with the first one. These steps are continued until a successive linear combination is no longer significant. Thus the purpose of canonical correlation is to explain the relationship of the two sets of variables, not to model the individual variables. CCA looks for the dimensions that account for the relationship and not for strength of the relationship. The canonical correlation squared gives the percent of variance in the dependent set explained by the independent set of variables along a given dimension. CCA involves eigen analysis and the matrix has the correlation of predictor and predictand (cross-dataset). The elements of this matrix is processed and then subjected to EOF analysis. EOF analysis reduces noise and overfitting, as the resulting components are orthogonal. CCA is then applied on these EOF modes, which gives the amplitude of few EOFs of the predictor and predictand fields.

5.3.1. Mathematical Steps Involved in CCA

Suppose $X_{s,t}$ and $Y_{r,t}$ denote mean removed data matrices corresponding to predictor and predictands. The space and time of these data fields are represented by s, r and t respectively where

s=1,2,…..,m
r=1,2,…..,n
t=1,2,……,k

EOF analysis on these data fields gives

$$X_{s,t}=E_{s,s}T_{s,t} \tag{120}$$

$$Y_{r,t}=E_{r,s}\,T_{r,t} \tag{121}$$

of which the $E_{s,s}$ and $E_{r,s}$ are the EOF modes and their corresponding time coefficients are given by $T_{s,t}$ and $T_{r,t}$ respectively. After eliminating unimportant modes, we have i EOF modes of predictor dataset and j EOF modes of predictand datasets. Thus i< m and j< n are the inputs for CCA analysis.

Now the canonical vectors (u, v) and their linear combinations $Z=u'T_{i,t}$ and $W=v'T_{j,t}$ can be calculated.

$$U' = \left[u_1', u_2', \ldots, u_q'\right] \tag{122}$$

$$V' = \left[v_1', v_2', \ldots, v_q'\right] \tag{123}$$

$$A_{q,q} = \begin{bmatrix} a_1 & & \\ & a_2 & 0 \\ & & . \\ 0 & & a_q \end{bmatrix}$$

where a_1, a_2, a_q are canonical correlations between Z and W and $a_1 \geq a_2 \geq \ldots\ldots \geq a_q$ and q is equal to i or j, whichever is smaller. Forecast equation at time t for a lead time l for the first j modes of predictand can then be constructed as

$$\hat{T}_{j,\hat{t}+1} = \left(V'\right)^{-1} A_{q,q} U' T_{i,t} \tag{124}$$

due to orthogonal property, eqns. (119) and (120) can be truncated as

$$T_{i,t} = \left(E_{s,i}\right) X_{s,t} \tag{125}$$

$$Y_{r,t} = E_{r,j} T_{j,t} \tag{126}$$

using eqns. (124) and (125), $T_{j,t+1}$ can be transformed to predictand values as

$$\hat{Y}_{r,t+1} = E_{r,j}(V')^{-1} A_{q,q} U'(E_{s,i}) X_{s,t} \tag{127}$$

An example, KOR7 of the application of canonical correlation analysis can be seen in Korecha and Barnston (2007), where the Ethiopian seasonal rainfall is a predicted using global SST field. CCA is performed on global SST (predictor) and Ethiopian JJAS standardised rainfall anomalies over individual stations (predictand) datasets for the period 1970-2002. Two predictors were used; May SST and JAS (July-September) SST. Using cross validation only two modes were selected and thus only two modes retained from CCA. The CCA gives patterns that are correlated both with the SST and the Ethiopian rainfall anomaly. Spatial maps resulted from CCA on (1) May SST and Ethiopian JJAS rainfall and (2) JAS SST and Ethiopian JJAS rainfall are shown in Figures 19 and 20 respectively. Canonical correlation is the correlation between the predictor and predictand temporal score and it gives the strength of the relationship between the two. For May SST, the Canonical Correlation is 0.55, and for JAS SST, it is 0.70, indicating a stronger relationship than for May SST. The inclusion of the second component did not improve the results and therefore discarded. The temporal components of the first mode of CCA were utilized in a simple regression model to predict the Ethiopian rainfall utilizing the temporal pattern of SST. The result of regression obtained for May SST and JAS SST with Ethiopian JJAS rainfall is shown in Figure 21. The simple linear regression is fitted over 1970–96 and tested over 1997-2002. The skills are expressed as the percentages of total variance explained. This is about 25% and 49% of all-Ethiopian JJAS rainfall for May SST and JAS SST respectively.

The CCA temporal modes utilised here to develop a linear regression model can be made use to predict the rainfall over a space (region) by fitting the linear regression equation at each spatial grid. The basic requirement for such prediction is that the dependent variable should be distributed over a space than at individual stations.

6. Forecast Verification

The forecasts obtained from different models are evaluated and this step is commonly known as forecast verification.

6.1. Evaluation and Correction

The forecasts are verified using different skill scores. The skill scores used vary from one method to the other. The aim of this section is to discuss the different statistics used to assess the different statistical models.

6.1.1. Evaluation Statistics of Time Fields

Some of the statistics given in this section are discussed in Nicholls 1984, Hastenrath 1987.

6.1.1.1. The Mean Systematic Error or Bias

$$\mathrm{BIAS} = \sum_{i=1}^{n} (\hat{X}_i - X_i)/n \tag{128}$$

where $\hat{X}$ is the forecast value, X_i is the observed value at time i and n, the number of observations.

6.1.1.2. Absolute Error (ABSE)

This quantity is slightly different from bias as it takes the absolute value of the difference between forecast and observed values.

$$\mathrm{ABSE} = \sum_{i=1}^{n} \left|\hat{X}_i - X_i\right|/n \tag{129}$$

6.1.1.3. Mean Absolute Error (MAE)

This is a simple way to avoid the compensation of positive and negative forecast errors. This is defined as the mean of absolute values of the individual forecast errors. This is given by the statistic

$$\mathrm{MAE} = \sum_{i=1}^{n} \left|\hat{X}_i - X_i\right|/n \tag{130}$$

The statistic is useful as it is more resistant to outliers.

6.1.1.4. Mean Squared Error (MSE)

MSE is estimated as

$$MSE = \sum_{i=1}^{n} (\hat{X}_i - X_i)^2 / n \qquad (131)$$

MSE is more sensitive to forecast errors than MAE, due to the square term in the statistic. This statistic is excessively sensitive to outlier errors than being a representative of the forecasts as a whole.

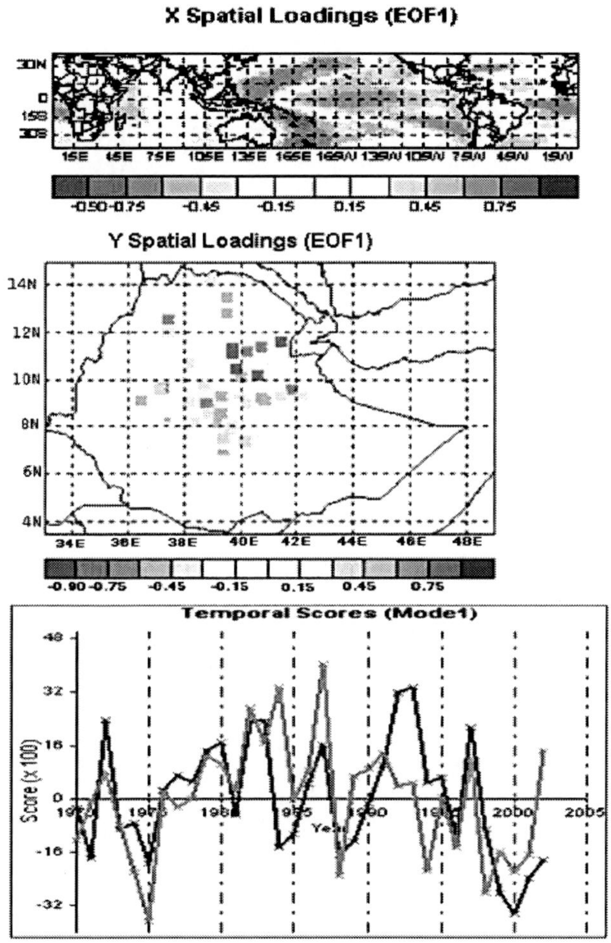

Figure 19. The Spatial loadings of the first CCA mode (called EOF1) for the prediction of Ethiopian JJAS station rainfall (Y spatial loadings), based on May SST (X spatial loadings). The time series of the temporal scores of the predictor SSTs (gray line) in predicting JJAS rainfall (black line) for this CCA mode. Based on 1970–2002 data. (Adapted from Korecha and Barnston 2007).

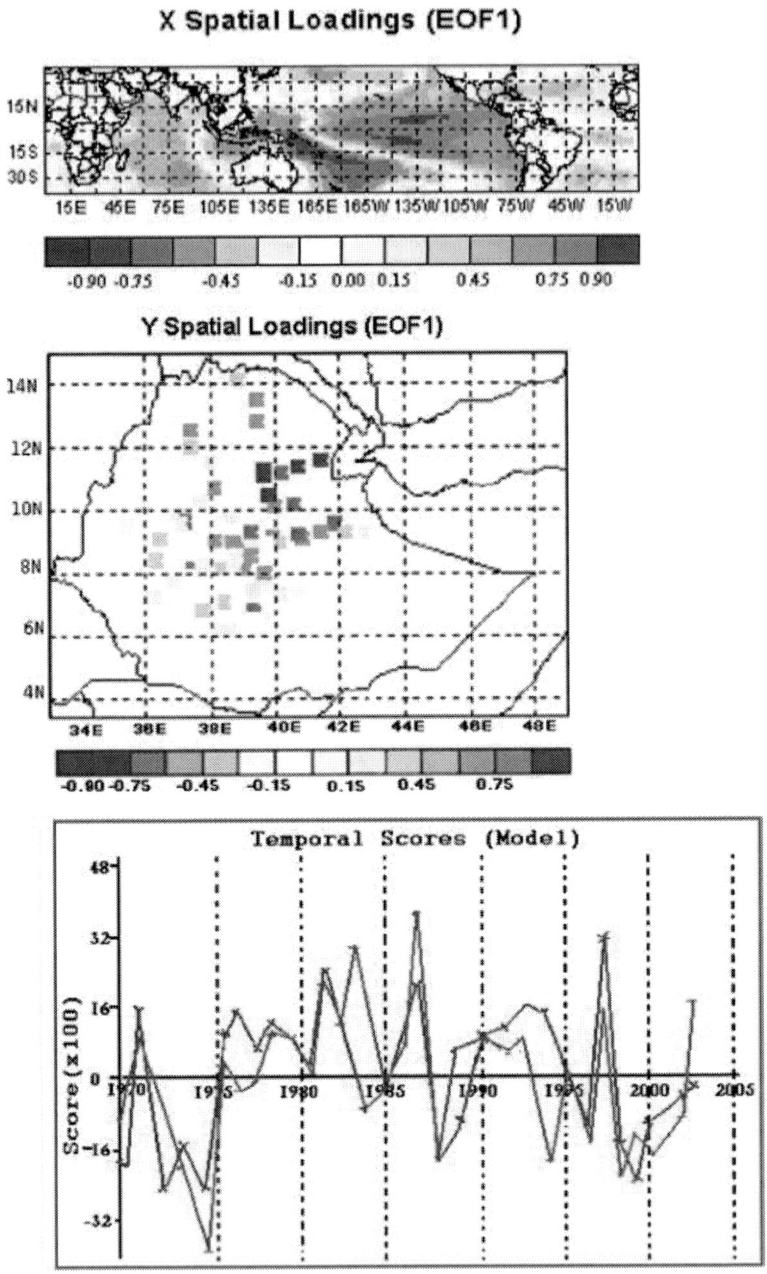

Figure 20. Same as Figure 19 but for JAS SST (Adapted from Korecha and Barnston 2007).

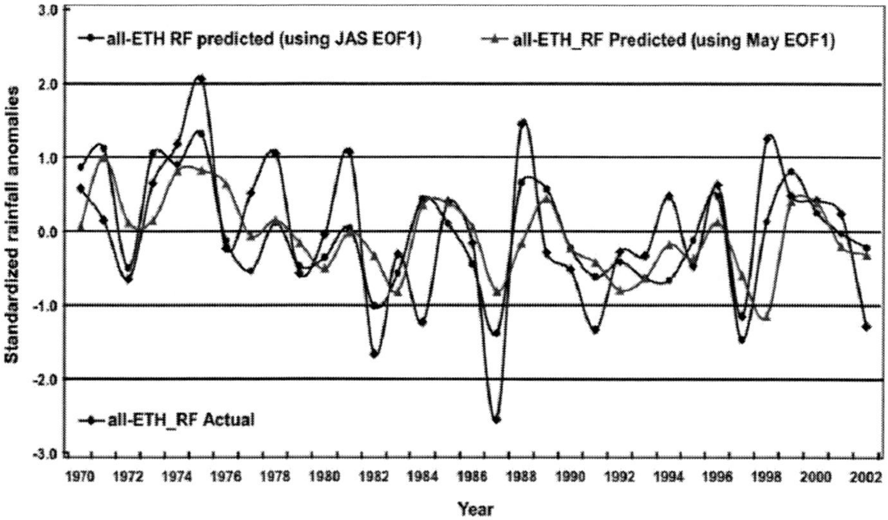

Figure 21. Model predicted (lines with circles and triangles) and observed (line with diamonds) all-Ethiopian JJAS standardized rainfall anomalies, based on May and JAS CCA mode 1. (Adapted from Korecha and Barnston 2007).

6.1.1.5. Root Mean Square Error (RMSE)

The square root of MSE gives RMSE,

$$\mathrm{RMSE} = \left[\sum_{i=1}^{n} (\hat{X}_i - X_i)^2 / n \right]^{\frac{1}{2}} \tag{132}$$

6.1.1.6. MSE Skill Score (MSESS)

MSESS is proposed by Murphy and Epstein (1989). This is used to judge the RMSE score by comparing it with the RMSE of a low-skill forecast. This is defined as

$$\mathrm{MSESS} = 1 - \frac{\mathrm{MSE}}{\mathrm{MSE}_{clim}} \tag{133}$$

where MSE_{clim} is the MSE for climatological mean forecasts. Climatological forecast requires no model development but is based on climatology of the past observation. When the mean bias of the forecasts is zero, this quantity equals to the correlation between the forecasts and observations. MSESS can have a maximum value of one indicating a perfect forecast and zero corresponding to a forecast equivalent to climatological forecast. A negative value implies the model is poorer than a climatological prediction.

6.1.2. Evaluation Statistics of Probabilistic and Categorical Forecasts

6.1.2.1. Relative Operating Characteristic (ROC)

ROC is a graph of the hit rate (Y-axis) against false alarm rate (X-axis) for different decision thresholds. This is useful in obtaining quantitative estimates of the probabilities of forecast outcomes.

6.1.2.2. Hit Score (HS)

HS gives the proportion of correct forecasts and is calculated as the ratio of the correct forecasts to the total number of forecasts.

6.1.2.3. Heidke Skill Score (HSS)

HSS is the adjusted HS to account for the proportion of forecasts that would be correct by chance in the absence of skill.

$$HSS = [(3/2) \times HS] - 1/2 \qquad (134)$$

HSS should be more than 0.33 for a good model ie 1/3rd of the forecasts are correct.

6.1.2.4. False Alarm Rate (FAR)

It is measured as the percentage of the total forecast in that category which is opposite to that of observed category. FAR for a below (above) normal forecast is the percentage of the forecasts made in that category for which observed category was above (below) normal.

6.1.2.5. Probability of Detection (POD)

POD explains the goodness of the model to detect a pre-defined category. POD for below normal category is the probability of the model forecasting below normal when below normal rainfall was observed.

6.1.3. Verification for Spatial Fields

The measures used in spatial domain is almost the same for time fields. The only difference is that the statistic is applied over the spatial field. Thus the statistics such as Bias, MAE, MSE are also used, but the difference is that these statistics are applied on spatial fields. A statistic, which is commonly used and not discussed earlier, is the Anomaly Correlation Coefficient.

6.1.3.1. Anonamly Correlation Coefficient (ACC)

It is a widely used measure to assess the performance over a spatial field. It is defined as

$$\mathrm{ACC} = \frac{\sum_{i=1}^{n}(\hat{X}_i' - \overline{\hat{X}'})(X_i' - \overline{X'})}{n s_{\hat{X}'} s_{X'}} \tag{135}$$

$$X_i' = X_i - X_{ci} \tag{136}$$

$$\hat{X}_i' = \hat{X}_i - X_{ci} \tag{137}$$

where $\hat{X}$ is the forecast field, X the observed field and X_c the climatological field. More on this topic can be read from Murphy and Winkler (1987) and Wilks (1995)

7. Conclusion and Concluding Remarks

This chapter has described various techniques and methods that are used in empirical rainfall forecasting. The basic tools and the various qualitative/quantitative forecast models are discussed. The different models to handle the forecast of temporal and spatial fields are detailed with examples from atmospheric sciences. Even though the various empirical models are available to forecast rainfall, the linear and non-linear interactions and the highly variable nature makes the rainfall forecasting complex. This is reflected by the inadequacy of the empirical models. Although this inadequacy has been solved to some extent by the introduction of numerical models, but its efficiency to forecast rainfall is still a matter of debate.

Forecasts are made at different time intervals – short, medium and long-range. Whatever the intervals at which the forecasts are delivered, it is of use only when the forecasts are comparable to reality and disseminated on time. Short-range forecasts are very useful because the results can be extrapolated and the seasonality also can be captured. Therefore the statistical forecasts are useful in planning, but these predictions are altered when some uncertain things takes place, which makes the forecasts invalid. The medium range weather forecasts are useful to plan for a severe synoptic event (eg: cyclone) that is to occur in a week's time. Whereas in long-range predictions long-term trends are important and such forecasts warn in advance about any adverse situation arising after a longer time. A failure/excess in the monsoon, predicted much earlier can help the government to plan for the water requirement and other agricultural needs of the country. The socio-economic issues can be settled and thus help to foresee the future shortcomings/excess by adapting ways to handle the forthcoming change. The forecasts delivered will be useful only if it is given at a right time and taken into consideration seriously. The use of the forecasts depends on the accuracy, measures taken to foresee the situation and the act of the users on these forecasts. If there is no planning, then there is no use of the forecasting even though it is 100% accurate.

Issuing a forecast is a challenging task, as it needs to satisfy the requirement to a higher degree. The difficulty in forecasting lies in each and every step like finding suitable predictors, framing a model and disseminating accurate/successful forecasts. The complexity in finding suitable predictors exists, as the rainfall is highly variable in space and time. The topographical changes such as the presence of mountains-valleys and land-sea contrasts makes forecasting a complicating job with the addition of oscillations and periodicities in the time series. Even if these variabilities are known, it is difficult to make indices that account for all these, as all such relationships are not linear but highly non-linear. Developing a non-linear model is a solution to account for the non-linear relationship but finding suitable predictors are mostly based on linear assumptions. All such factors contribute to the inefficiency of the model. The forecasts from such statistical models are not sufficient to give accurate forecasts. These are some of the challenges that have to be overcome while forecasting. These inadequacies get even worse with the incapability of the forecaster. Any forecast distributed is effective only when it is utilized for decision-making. This purpose is not served, if it is not delivered well before the occurrence of the event, if there is no response from the authorities or if it is not as per the requirement. The lack of good forecasting methods, knowledge of advanced techniques and less improvement add to the forecasting problems. These are some of the problems that usually arise while disseminating forecasts. Thus, the inaccurate forecasts are in short, the result of unsophisticated methods and untrained forecasters. But training the forecasters and utilizing better and more skillful methods to capture the uncertainties in forecasting can resolve these problems to some extent.

Even though statistical models are used at a greater rate than the numerical approach, nevertheless they have shortcomings that at times new models with new parameters need to be developed especially when the predictors loose their strength of the relationship with the predictand. Thus the development of new empirical models with sufficient skill is still required in order to satisfy societal needs.

References

Akaike, H., (1974) A new look at statistical model identification, *IEEE transactions on automatic control,* AC-19, 716-723.

Anderson, R.L., (1942). Distribution of the serial correlation coefficient, *Annals of Mathematical Statistics,* 13, 1-13.

Barlett, M.S., (1946) On the theoretical specification of sampling properties of autocorrelated time series, *Journal of the Royal Statistical Society,* Series B., 8, 27.

Bishop, C.M., (1995) *Neural networks for pattern recognition.* Clarendon Press, Oxford, pp 135–136.

Blanford, H.H., (1884) On the connection of Himalayan snowfall and seasons of drought in India. *Proc. R. Soc. London,* 37, 3-22.

Box, G.E.P., Jenkins, G.M., (1970) *Time Series analysis: Forecasting and control,* San Francisco: Holden-Day.

Chan, J.C.L., Shi, J., (1999) Prediction of the summer monsoon rainfall over South China. *Int. J. Climatol* 19:1255–1265

Chan, J.C.L., Shi, J., Lam, C., (1998) Seasonal forecasting of tropical cyclone activity over the Western North Pacific and the South China Sea. *J Weather Forecast* 13:997–1004

Fuller, W.A., (1976). *Introduction to statistical time series,* New York: John Wiley and Sons.

Dickey, D.A., Bell, W.R., Miller R.B., (1986). Unit roots in time series models: tests and implications, *American Statistician,* 40, 12-26.

Draper, N.R., Smith, H., (1981) *Applied Regression analysis,* 2nd ed., New York: John Wiley and Sons.

Fox, J., (1991) *Regression Diagnostics: An Introduction. (Sage University Paper series on Quantitative Applications in the Social Sciences,* series no. 07-079). Newbury Park, CA: Sage.

Friedman, J.H., (1984) *A variable span smoother. Department of Statistics,* Stanford University. Technical report LCSS

Friedman, J.H., Stuetzle, W., (1981) Projection pursuit regression. *J Am Statist Assoc* 76:817–823

Friedman, J.H., Tukey, J.W., (1974) A projection pursuit algorithm for exploratory data analysis. *IEEE Trans Comput* 23:881–890

Glahn, H.R., (1968) Canonical correlation and its relationship to discriminant analysis and multiple regression. *J. Atmos. Sci.,*25, 23–31.

Gowariker, V., Thapliyal, V., Kulshrestha, S.M., Mandal, G.S., Sen Roy, N., Sikka, D.R., (1991) A power regression model for tong range forecast of southwest monsoon rainfall over India. *Mausam* 42: 125-130

Guiot, J., Berger, A.L., Munaut, A.V., (1982) *Response functions*. In: Hughes, M.K.; Kelly, P.M., Pilcher, J.R., LaMarche, V.C., eds. Climate from tree rings. Cambridge, UK: Cambridge University Press: 38-45.

Hastenrath, (1987) On the prediction of India monsoon rainfall anomalies. *Journal of Climate and Applied* Meteorology 26:847-857

Hotelling, H., (1935) The most predictable criterion. *Journal of Educational Psychology,* 26, 139-142.

Huber, P., (1985) *Projection pursuit. Ann Stat* 13:435–475

Jolliffe, I.T., (2002) Principal Component Analysis. 2nd ed. Springer, 487 pp.

Korecha, Diriba, Anthony, G., *Barnston*, (*2007*) Predictability of June–September Rainfall in Ethiopia. *Mon. Wea. Rev.,* 135, 628–650

Krishna Kumar, K., Soman, M.K., Rupa Kumar, K., (1995) Seasonal forecasting of Indian summer monsoon rainfall. *Weather* 50: 449–467.

Li, Y., Smith, I., (2008) A Statistical Downscaling Model for Southern Australia Winter Rainfall. *Journal of Climate.* 22:1142-1158.

Loehlin, J.C., (1992) Latent variable models: an introduction to factor, path, and structural analysis. Hillsdale, N.J., *Lawrence Erlbaum Associates*. 292 p.

Makridakis, S., Wheelwright, S.C., Hyndman, R.B., (1998) Forecasting: methods and applications. *John Wiley and Sons.* 241-310 pp.

McCuen, R.H., (1985) *Statistical methods, for Engineers.* New Jersey: Prentice Hall, 439 pp.

Momani, P.E.N.M., (2009) Time Series Analysis Model for rainfall data in Jordan: case study for using Time series analysis. *American journal of Environmental sciences* 5: 599-604.

Murphy, A.H., and Winkler, R.L., (1987) A general framework for forecast verification. *Mon. Weather Rev.,* 115, 1330–1338.

Myers, R.H., (1986) *Classical and modern regression with applications. Boston*: Duxbury Press. 359 pp.

Nayagam, L.R., Janardanan, R., and Mohan, H.S.R., (2007) An empirical model for the seasonal prediction of southwest monsoon rainfall over

Kerala, a meteorological subdivision of India. *International Journal of Climatology,* 28(6), 823-831.

Neter, J., Wasserman, W., Kutner, M.H., (1990) Applied Linear Statistical Models. 3d ed. Irwin, 1181 pp.

Nicholls, N., (1984) The stability of empirical long-range forecast techniques: A case study. *Journal of Climate and Applied Meteorology* 23:143-147.

Quenouille, M.H., (1949) The joint distribution of serial correlation coefficients, *Annals of Mathematical Statistics,* 20, 561-571.

Rajeevan, M., Guhathakurta, P., Thapliyal, V., (2000), New Models for Long range forecasts of summer monsoon rainfall over NW India and Peninsular India, *Met and Atmos. Phys.,* 73, 211-225.

Rajeevan, M., Pai, D.S., Dikshit, S.K., Kelkar, R.R., (2004) IMD's new operational models for longrange forecast of southwest monsoon rainfall over India and their verification for 2003, *Current Science* 86: 422-431

Rajeevan, M., Pai, D.S., Anil Kumar, R., Lal, B. (2006) *New statistical models for long-range forecasting of southwest monsoon rainfall over India.* 28: 813-828.

Roosen, C.B., Hastie, T.J. (1994) Automatic smoothing spline projection pursuit. *J. Comput. Graph. Stat.* 3:235–248.

Rumelhart, D.E., Hinton, G.E., Willimas, R.J., (1986a) *Parallel Distributed Processing,* Vol 1, Cambridge MA: MIT Press, pp.318-362.

Rumelhart, D.E., Hinton, G.E., Willimas, R.J., (1986b) Learning representations by back propagating errors. *Nature,* 323, 533-536.

Sahai, A.K., Chattopadhyay, R., Goswami, B.N., (2008) *A SST based large multi-model ensemble forecasting system for Indian summer monsoon rainfall, Geo. Res. Let.*, 35, L19705,DOI 10.1029/2008/GL035461. ISSN 0094-8276.

Stone, M., (1974) Cross-validatory choice and assessment of statistical predictions. *J. Roy. Stat. Soc.,* 36, 111–147.

Thapliyal, V., (1982) Stochastic Dynamic Model for long range forecasting of summer monsoon rainfall in Peninsular India, *Mausam, 33*, 399-404.

Thapliyal, V., (1987) Prediction of Indian monsoon variability evaluation and prospects including development of a new model. In: D. Ye., C Fu; J. Chano, and M. Yoshino, (Eds.) *Climate of China and global climate,* China Ocean Press, 397- 416 pp.

Venkatesan, C., Raskar, S.D., Tambe, S.S., Kulkarni, B.D., Keshavamury, R.N., (1997) Prediction of All India Summer Monsoon Rainfall Using Error-back-propagaion Neural Networks Meteorol. *Atmos. Phys.* 62, 225-240.

Walker, G.T., (1923) Correlations in seasonal variations of weather, VIII. A preliminary study of World Weather I. *Memoirs of India Meteorological Department,* 24, 75-131.

Walker, G.T., (1924) Correlations in seasonal variations of weather, IX, A further study of world weather (World weather II). *Memoirs of India Meteorological Department,* 24, 275-332.

Wilks, D.S., (1995) *Statistical Methods in the Atmospheric Sciences: an Introduction, International Geophysics Series* 59: Academic Press, 464 pp.

In: Rainfall Forecasting
Editor: T.S.W. Wong, pp. 129-160
ISBN: 978-1-61942-134-9

Chapter 3

THE CHALLENGE OF WARM-SEASON CONVECTIVE PRECIPITATION FORECASTING

William A. Gallus, Jr.
Department of Geological and Atmospheric Sciences
Iowa State University, Ames, IA, 50011, U.S.A.

ABSTRACT

Deep moist convection presents numerous challenges for precipitation forecasting, as evidenced by objective skill scores for warm season quantitative precipitation forecasts (QPF) that are far lower than those for the cold season. Convective precipitation occurs on small-scales with large gradients in precipitation amounts and short lifetimes marked by high intensities, allowing even small spatial and temporal errors to result in poor objective skill scores and large negative impacts on some users of the forecasts. To make matters worse, convection is often forced by small-scale phenomena as well, such as convective outflow boundaries, soil moisture gradients, or even horizontal boundary layer rolls, all of which are rarely forecasted well by numerical weather prediction systems. Coarser grid models may use convective parameterizations which introduce large errors; even finer grid models suffer from deficiencies in microphysical and planetary boundary layer parameterizations.

Because of the many sources of error affecting convective precipitation forecasts, both ensemble techniques and postprocessing procedures are increasingly used to improve the forecasts. Although most ensembles have been found to work best when using perturbed initial and lateral boundary conditions to generate members, convective rainfall forecasting may be an exception where the use of mixed physical parameterizations and models yields an ensemble with increased spread and skill for better reliability and resolution. As with other weather parameters, rainfall forecasts based on an ensemble mean are more skillful than forecasts from any individual member; however, the ensemble mean smooths small-scale high-amplitude features too much, usually resulting in low biases for heavy amounts and high biases for low amounts. A probability-matching approach has been found to yield better skill and bias for the ensemble mean. Post-processing can help to improve these rainfall forecasts and may include bias-correction and neighborhood approaches.

Because traditional objective skill measures double penalize high resolution forecasts often used for convective forecasting and may not match impressions from human forecasters, convective rainfall forecasts are increasingly evaluated using new verification techniques such as object-based approaches. Such approaches can help modelers to understand deficiencies in the model, but the more complex output from the techniques may be more difficult for forecasters to understand.

Keywords: Moist Convection, Warm Season, Precipitation Forecasting, Parameterizations, Numerical Weather Prediction, Mesoscale Meteorology, Thunderstorms, Ensemble Forecasting, Forecast Verification, Model Initialization, Rainfall, Microphysics, Post-processing.

INTRODUCTION

Although skill for quantitative precipitation forecasting (QPF) has improved substantially in recent decades, skill has remained much lower during the warm season when convective precipitation contributes a large portion of the rainfall (Figures 1 and 2). In addition, improvement in skill has been much slower for warm season QPF in the last 20 years or so than for cold season QPF, resulting in an increasing gap in skill between these seasons. A comparison of Figures 1 and 2 shows that threat scores, one objective verification measure, for 24h QPF during the period February 2010 through February 2011 in most cool season months exceeded the maximum values that occurred during the 1984-1993 period (Olson et al. 1995), but scores during the summer months have improved noticeably less over this time period. This lack of improvement for the warm season may at first be surprising since increases in computer power that have occurred in recent decades have allowed models to be run with much finer grid spacings, and warm season precipitation, often from convection, generally occurs on smaller spatial and temporal scales than precipitation during the cool season. Thus, one might expect summer QPF to benefit more from the refinement in resolution. Figure 2 also shows that the human forecasters (denoted HPC for Hydrometeorological Prediction Center) are able to improve upon guidance taken directly from one numerical weather prediction model (GFS; Global Forecast System), but again, their improvement in the warm season is no better than it is at other months of the year. As further evidence of the problems with warm season QPF, Wilson and Roberts (2006) found that 10 km Rapid Update Cycle (RUC) 3-h forecasts of precipitation initiation during the warm season in the International H_2O Project (IHOP) were correct only 44% of the time, even allowing for a tolerance of 250 km in space and 3 h in time.

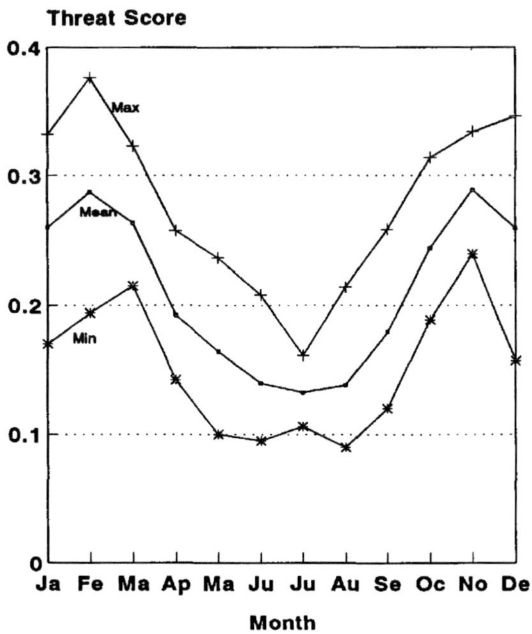

Figure 1. Mean, minimum, and maximum threat score for NCEP forecasters by month over period 1984-1993 for 1.0 inch threshold during day 1 (from Olson et al. 1995, © American Meteorological Society. Reprinted with permission).

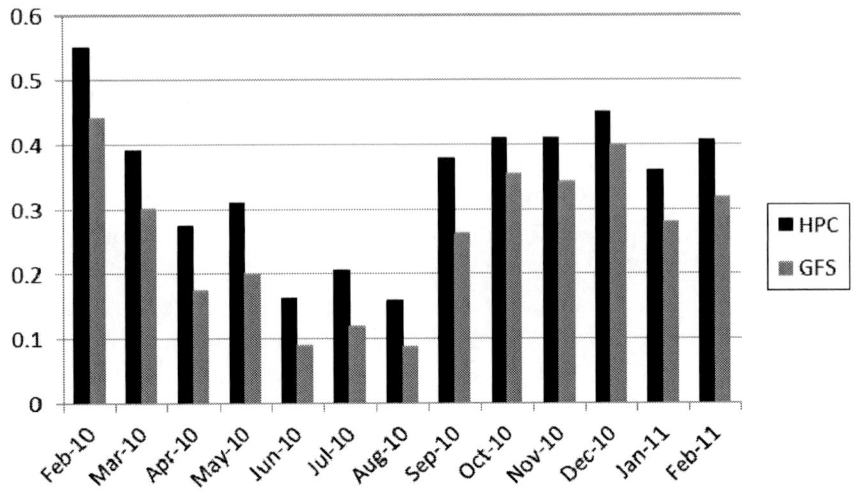

Figure 2. Threat scores for NCEP's Hydrometeorological Prediction Center and the GFS model as a function of month during the period Feb 2011 – Feb. 2011 for a 1.0 inch precipitation threshold during the day 1 period.

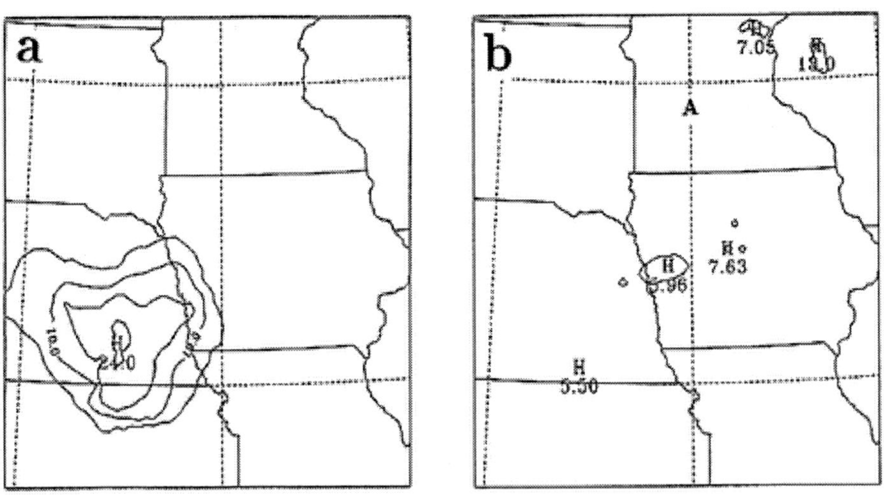

Figure 3. Accumulated precipitation (mm) in 12-h period (1200–0000 UTC 16-17 Jul 1996) from the 22-km Eta Model with the (a) BMJ and (b) KF convective parameterizations (from Gallus 1999, © American Meteorological Society. Reprinted with permission). Contour interval is 5 mm.

The lack of skill and similar lack of improvement in warm season QPF are due to a variety of reasons which are explained in this chapter. Many of these reasons are related to the small-scale nature of convective precipitation and the forcing mechanisms that cause it; however, problems inherent with traditional verification of precipitation also influence scores such as those shown in Figures 1 and 2, potentially making the problem seem worse than it is. Current methods designed to improve warm season QPF, along with verification techniques that yield results better matching subjective impressions of skill for these forecasts, will also be examined.

Difficulties in Forecasting Warm Season Precipitation

In many parts of the world, such as the agriculturally-important central Plains and Midwestern United States, the warm season supplies the majority of annual rainfall, often in the form of intense thunderstorms that can produce rainfall rates exceeding 10 cm h^{-1} with gradients in accumulated rainfall as large as 10 cm $(10\ km)^{-1}$. These events often lead to dangerous flash flooding that claims numerous lives each year (Doswell et al. 1996). In addition, mesoscale convective systems (MCSs; see Houze 2004 for thorough description), organized clusters of individual convective elements often accompanied by larger areas of lighter stratiform rainfall, are responsible for most of the warm season rainfall in some of these areas (Fritsch et al. 1986), and MCSs reach maximum intensity and areal coverage during the night, a time when the surface-based instability usually thought to fuel most thunderstorms is on the decrease. These MCSs are primarily forced by mechanisms that are only discernible aloft (Maddox 1983, Cotton et al. 1989), such as strong convergence at the nose of the low-level jet (see Stensrud 1996 for review). Whether the convection occurs as isolated cells during the daytime, or in an organized nocturnal MCS, forecasting is difficult because (1) the individual convective elements occur on scales too small to be adequately resolved by even the higher-resolution operational models used today, and (2) the mechanisms resulting in the convection are also of small-enough scale so that they may be inadequately resolved in the models (Kain and Fritsch 1992, Stensrud and Fritsch 1994). The former difficulty results in the need for parameterizations of important physical processes in the models. The difficulties related to both parameterizations and to the small-scale forcing mechanisms are discussed in the subsections to follow.

Role of Parameterizations

Bryan et al. (2003) has shown that to truly resolve convective cores, a horizontal grid spacing of 100 m or less would be needed in numerical weather prediction models. Such fine grid spacings are prohibitively expensive for use in operational models today and will likely remain so for many years to come. Thus, even the finest horizontal grid spacings used quasi-operationally today, on the order of 1-4 km within regional or limited area models, at best can be termed "convection-allowing" as opposed to "convection-resolving" (Kain et al. 2006). Nonetheless, Weisman et al. (1997) concluded that a horizontal grid spacing of 4 km was sufficiently small to resolve vertical motions well enough to result in reasonably good simulations of convective systems, and such grid spacings are now being used increasingly often in operational and quasi-operational settings.

Traditionally, convective parameterizations were used to account for the inability of models to explicitly resolve convective impacts on the atmosphere (a thorough discussion can be found in Stensrud 2007). Many studies have shown that warm season precipitation forecasts are significantly sensitive to the choice of the convective scheme (Figure 3) and tunable parameters within schemes, and no one convective scheme works best for all cases (e.g., Kuo et al. 1996; Wang and Seaman 1997; Spencer and Stensrud 1998, Gallus 1999). With the convection-allowing horizontal grid spacings used quasi-operationally today in limited area models, traditional convective parameterizations are not physically justifiable (Molinari and Dudek 1992), and are usually neglected. Nonetheless, some studies have suggested that forecast skill may improve when these schemes are used, even down to roughly 5 km grid spacing (Kain and Fritsch 1998). These studies suggest that some processes important in the initiation or evolution of convection are not adequately resolved with 4 or 5 km grid spacing, and models may need assistance through parameterization of at least a subset of the processes represented in traditional parameterizations. One common problem found in some fine-grid simulations, for instance, is that precipitation development is delayed when a convective scheme is not used (e.g Molinari and Dudek 1986, Lean et al. 2008). Bukovsky and Kain (2006) have also pointed out that realistic upscale growth of convective effects cannot occur when convective parameterizations are used because the parameterizations act independently in individual model grid columns.

Even when convective parameterizations are not needed, strong sensitivity of QPF to parameterizations of other physical processes, such as microphysics and turbulence, particularly in the planetary boundary layer (PBL), have been noted (Jankov et al. 2005). Figure 4 (from Jankov et al. 2005) shows how correspondence ratios (defined as the ratio of intersection of rain area in two model runs to the union of rain area in the same model runs) and correlation coefficient (r^2) changed, on average, as convective, microphysical, and boundary layer parameterizations were changed in 12 km grid spacing Weather Research and Forecasting (WRF) model simulations of 8 convective system events occurring during the International H_2O Project. Sensitivity was greatest to the convective scheme (lowest values), with microphysical and boundary layer schemes playing somewhat similar secondary, though still important, roles.

When a convective parameterization is not used, all precipitation in the model is produced by the microphysical scheme. Commonly used microphysical schemes today include both liquid and solid hydrometeors, with as many as six different hydrometeor classes (e.g., rain water, cloud water, water vapor, cloud ice, snow, graupel or hail) used by some. Traditionally most microphysical schemes have been based on an assumed drop or particle size distribution, and are known as bulk schemes. These schemes in the past usually predicted just one moment of the distribution, most often the mixing ratio, but double moment schemes that also predict number concentration, or even triple moment that might predict radar reflectivity as well (e.g. Milbrandt and Yau 2005) are becoming more common. In recent years, some bin or spectral schemes also have been created which allow different-sized hydrometeors to exist at each model grid point (e.g., Lynn and Khain 2007).These bin schemes greatly increase computational costs over bulk schemes, and generally cannot be run operationally at the present time.

Within microphysics schemes, numerous assumptions regarding tunable parameters introduce error into forecasts. Such parameters may include those related to the drop size distribution, such as intercept or hydrometeor density; fall speed, including diameters and air resistance; and formulations of microphysical processes. Gilmore et al. (2004) found that simulations of supercell thunderstorms varied as changes were made to hydrometeor densities and slope intercept. Potter (1991) found that fallspeeds of snow and graupel were very sensitive to the definition of particle size. Colle and Mass (2000) showed how changes in fall speed of snow affect the distance over which snow is advected and hence the location of precipitation. This advection could be critically important in the simulation of MCSs where hydrometeor advection is known to play a major role in development of stratiform rain regions (e.g. Johnson and Hamilton 1988, Gallus and Johnson 1991).

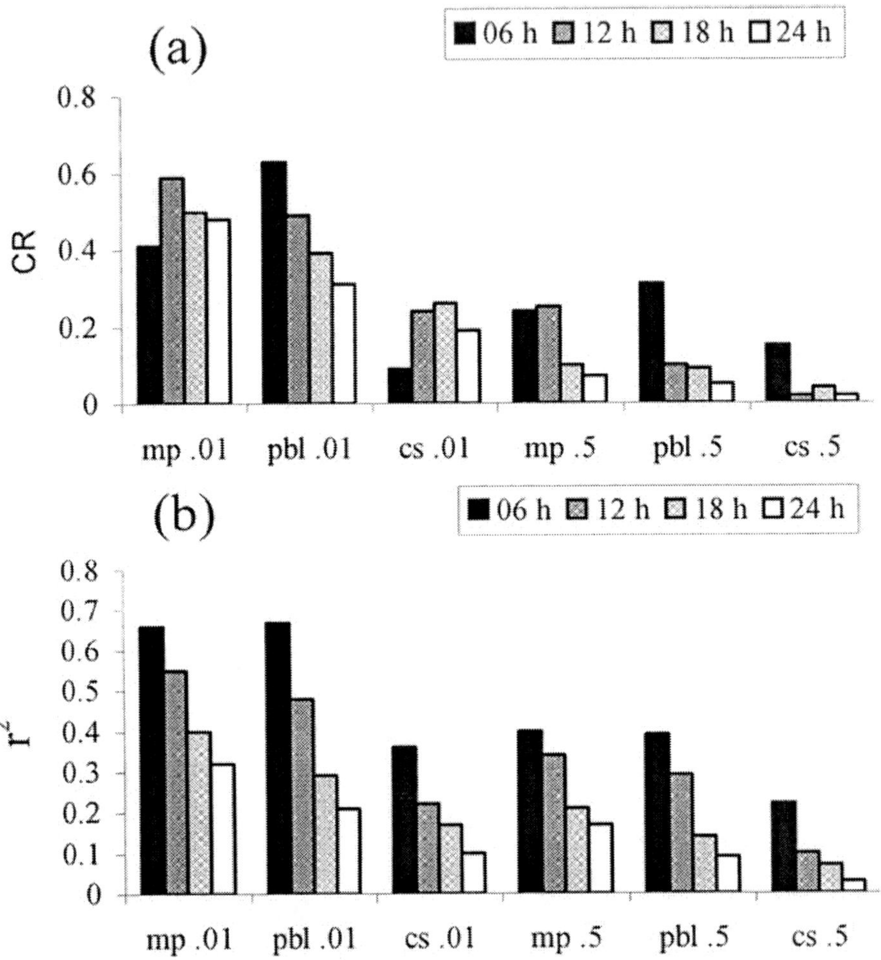

Figure 4. Values of (a) Correspondence Ratio (CR) and (b) r^2 for changes in microphysical (mp), PBL (pbl), and convective schemes (cs) within the WRF model. Results are presented for the two thresholds indicated along the abscissa (0.01 and 0.5 in.) and for the four 6-hourly periods ending at the times indicated in the legend (from Jankov et al. 2005, © American Meteorological Society. Reprinted with permission).

In addition, the relative coarseness of even convection-allowing grids introduces another source of error based on the threshold relative humidity at which a grid box is assumed to be saturated enough to allow processes like condensation to begin. In nature, condensation producing clouds and precipitation can begin if the saturation is present on very small scales, less than 1 km^2. With most weather forecasting models using coarser grid spacing, it is arguably unreasonable to require 100% relative humidity over the full grid box before allowing precipitation processes to begin. Most commonly used microphysics schemes use the 100% threshold, although until around 2006, the Ferrier scheme running in the operational Weather Research and Forecasting (WRF) model at the National Centers for Environmental Prediction in the Unites States and available to researchers across the world, used a scale-dependent threshold that decreased with increasing grid size. A lower threshold in some cases could allow convection to develop earlier, or perhaps in a region in which it otherwise would not with the 100% threshold.

With PBL schemes, large differences have been found in the daytime boundary layer development (e.g., Kain et al. 2006; Weisman et al. 2008) with some schemes like the Mellor-Yamada-Janjic (MYJ) which builds the boundary layer via direct mixing based on turbulence energy calculations having moister, cooler, and shallower boundary layers than others like the Yonsei University (YSU) which diagnoses an expected boundary layer depth and then mixes instantaneously through the entire boundary layer (Weisman et al. 2008). These differences in turn can affect the prediction of precipitation since even miniscule changes in the PBL temperature and moisture over the spatial scale of a few kilometers can determine whether or not convection will develop (Weckwerth et al. 1996).

Small-scale Nature of Forcing Mechanisms

Apart from issues that arise due to the small scale of convective elements, another serious problem in obtaining accurate forecasts is the small-scale of forcing mechanisms that result in convection through low-level convergence and lift (e.g. Byers and Braham 1949, Purdom 1976; Hill 1977, Ziegler et al. 1997, Crook and Klemp 2000, Moore et al. 2003). Many warm season convective events are believed to be triggered by features observable primarily at the surface, such as fronts (Shapiro 1981, 1982; Trier et al. 1991), outflow boundaries, trough lines, drylines (Schaefer 1986; Atkins et al. 1998, Ziegler et al. 1997), horizontal convective roles (e.g, Faller 1965, Lilly 1966), topographic and land use gradients, urban heat-islands (Baik et al. 2001), vegetation differences (Segal and Arritt 1992) and soil moisture gradients (Clark and Arritt 1995, Wilson and Roberts 2006). Although these mechanisms can initiate convection, Stensrud and Maddox (1988) identified cases when convergence boundaries did not initiate thunderstorms, even when they collided within conditionally unstable environments.

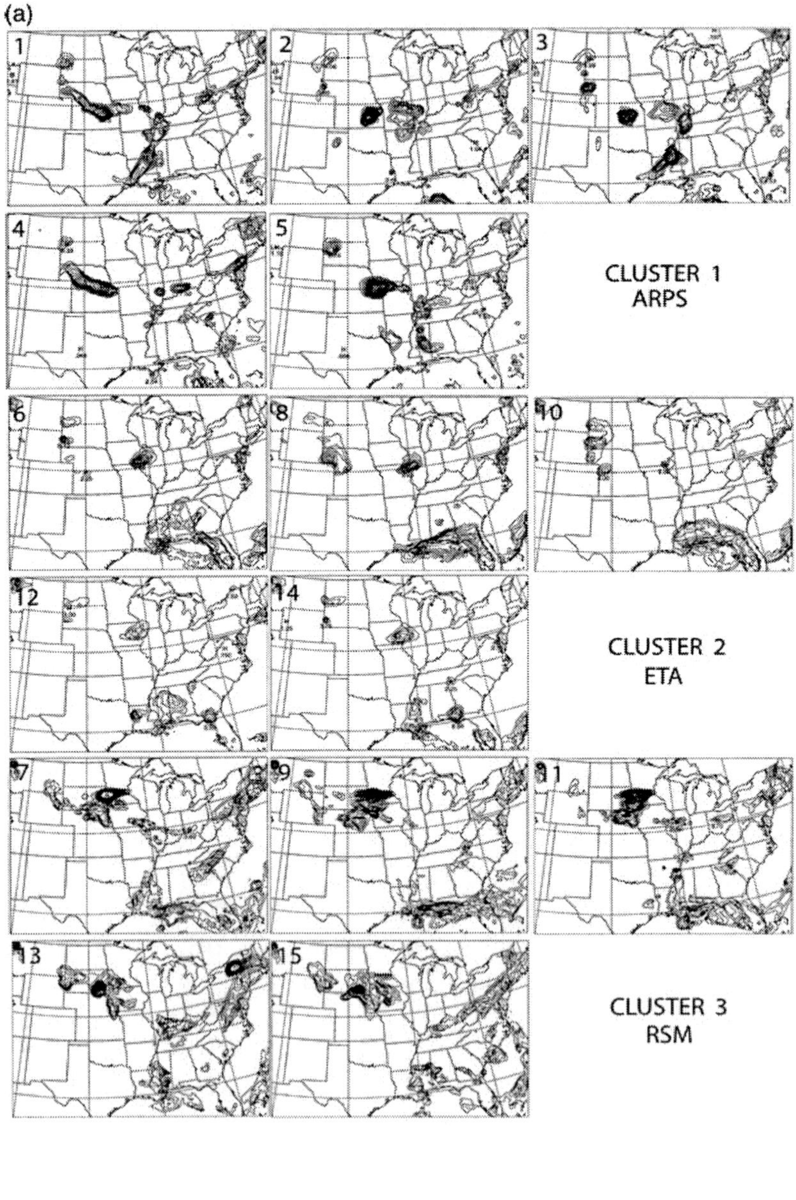

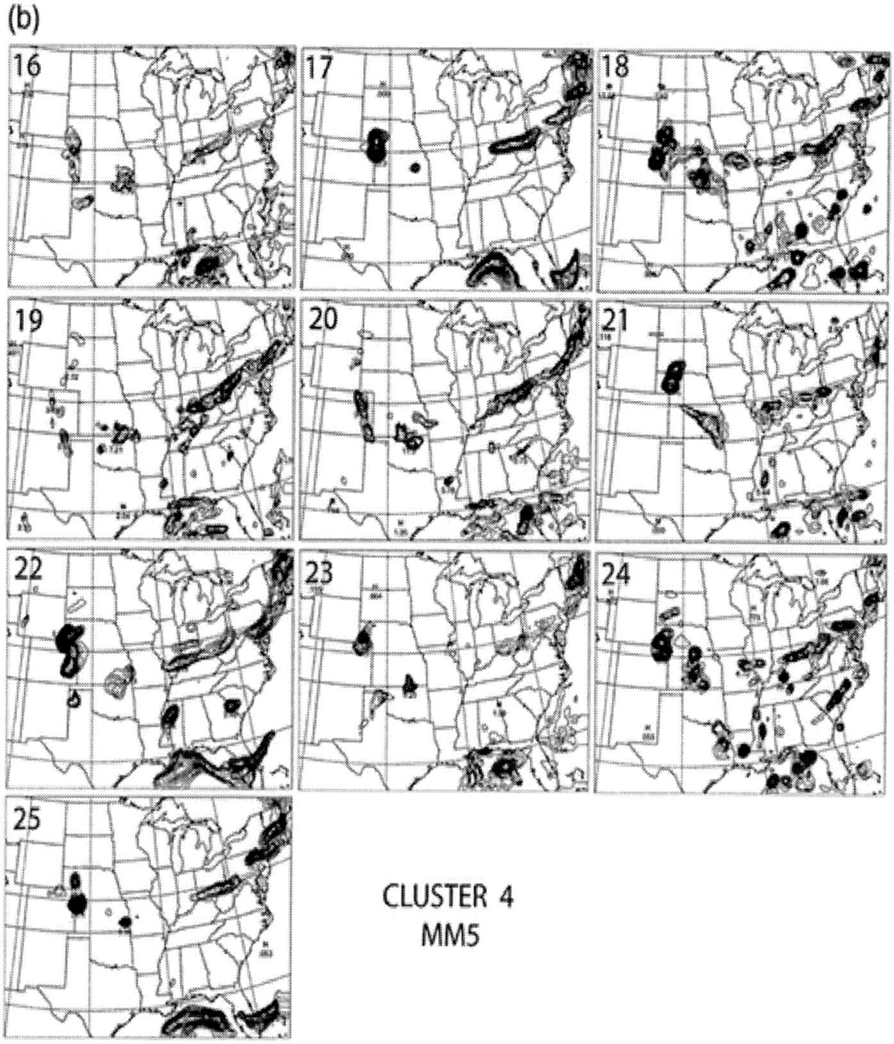

Figure 5. The 3-h accumulated precipitation (mm) valid at 30 h from 25 ensemble members grouped subjectively into four clusters (identified as the ARPS, ETA, RSM, and MM5 models). Numbers in the upper-left-hand corner indicate the different ensemble members. Isolines every 1 mm. From Alhamed et al. (2002), © American Meteorological Society. Reprinted with permission.

Considering soil moisture variations alone, Clark and Arritt (1995) noted that rainfall amounts were enhanced in locations that had nearly 100% moist vegetation coverage compared to locations that were covered by bare and dry soils. Gallus and Segal (2000) found that peak precipitation in simulations could be sensitive to even a 10% change in soil moisture, and the sensitivity to soil moisture varied depending on the convective parameterization used. In some situations, soil that was wetter led to a decrease in peak precipitation due to complicated feedbacks that developed as thunderstorms produced interacting outflow boundaries. In most of the world, soil moisture observations are much sparser than observations of typical atmospheric parameters such as temperature and moisture, meaning that many small-scale variations that might influence precipitation development are not resolved.

The problem is even worse for those systems often called "elevated convection" (LeMone and Meitin 1984; Colman 1990a, b; Ferrare et al. 1991; Wilson et al. 1994; Weckwerth et al. 1997) where a much sparser network of upper air observations makes it likely important mesoscale forcing mechanisms, such as internal gravity waves (IGWs; Weckwerth et al. 2008) or variations in moisture and temperature are not properly initialized in models. Crook (1996) showed that variations of 1 g kg^{-1} in planetary boundary layer moisture or 1°C in temperature were enough in numerical simulations to change whether thunderstorms developed or not . In addition, many convective parameterizations were designed with surface-based convection in mind, and may not work as well with elevated convection (e.g. Gallus 1999). Gallus et al. (2005) showed that one powerful derecho system initiating as elevated convection around sunrise that traveled over a roughly 1000 km path producing heavy rainfall and wind damage across the central United States was not simulated by any of over 20 simulations using a range of different models, physical parameterizations, and initial and boundary conditions. For such events, new methods of observing the atmosphere on small scales above the boundary layer may be needed before prediction can improve. More recently, during the 2011 National Oceanic and Atmospheric Hazardous Weather Testbed Spring Experiment, several cases were noted by the author where every one of the 15 members displayed to the participants of a 4 km ensemble run by the Center for the Analysis and Prediction of Storms failed to forecast important nocturnal convective systems that initiated within the first 12 hours of the forecasts. These ensemble members even made use of radar observations in their initializations. Such forecasting failures continue to show the need for better observing networks. Although assimilation of radar observations, as will be discussed in the next section, have been shown to improve forecasting skill on average, clearly for cases where no radar echo is present at initialization, the benefits may be more limited.

TECHNIQUES TO IMPROVE WARM SEASON FORECASTING

Although warm season rainfall forecasts will likely remain difficult for some time to come, slow progress is being made at improving the skill of these forecasts. Techniques that improve these forecasts are generally related to (1) improved initialization of fine-scale features, (2) improved models or modeling approaches (such as the use of ensembles), and (3) post-processing.

Improved Initialization of Simulations

Improved representation of mesoscale or storm-scale features has been investigated as one method to improve these forecasts (e.g., Xue et al. 2003). Stensrud and Fritsch (1994) and Stensrud et al. (1999b) investigated the impact of better initialization of cold pools on subsequent convective system rainfall and noted that some significant improvements occurred when special attention was paid to mesoscale details of the initialization. In addition, Gallus and Bresch (1997) showed that the addition of mesoscale observations valid through a deep atmospheric layer in the initialization could markedly improve a forecast; surface data alone had a limited impact. Gallus and Segal (2001), however, found only limited improvement when cold pools were initialized with the Stensrud et al. (1999b) technique, possibly because many of the 20 cases simulated were elevated nocturnal events where cold pools were limited in intensity by a near-ground stable layer. Gallus and Segal (2001) also found that the inclusion of surface mesonetwork data had little ability to improve the rainfall forecasts, even when a model's vertical diffusion scheme was used to modify the boundary layer based on the surface data. That study found a somewhat more significant improvement in rainfall skill scores if initialized relative humidity was altered over a deeper layer based on the position of radar echoes.

Many other studies also have found that using radar data in model initialization or assimilation systems can improve a forecast (e.g. Wang and Warner 1988, Takano and Segami 1993, Rogers et al. 2000). Improvements were noted both for techniques that were somewhat complicated, such as the determination of heating rate profiles from radar data (Wang and Warner 1988), and for relatively simple techniques, such as the activation of the convective scheme where radar echo was present (Rogers et al. 2000). More recently, extensive work has been performed on ways to assimilate both velocity and reflectivity information from Doppler radars. These approaches can include ensemble Kalman filter methods (e.g. Tong and Xue 2005) to computationally expensive 4-dimensional variational (4DVAR) data assimilation (e.g. Sun and Crook 1997). Xue et al. (2009) has found that simulation of radar data leads to substantial improvements in forecasts of convection, although the largest improvements are generally confined to the first 12 hours or so of a forecast.

Modeling Improvements

Another path toward improved warm season QPF is to improve the models themselves. As a sign of the extensive research in this area, an increasing number of physical parameterizations are made available to forecasters each year. For instance, in the research version of the WRF model, the number of PBL parameterizations available in 2011 is now 11, compared to only 4 just five years earlier. Among local closure PBL schemes, increasingly higher level (and order) versions have been added in recent years (e.g. Nakanishi and Niino 2006). Similarly, for microphysics, double moment schemes recently have been made easily available to the modeling community (e.g., Milbrandt double moment, Milbrandt and Yau 2005; Morrison double moment, Morrison et al. 2009), and some studies have examined even higher moment schemes (Milbrandt and Yau 2005) or bin approaches (Lynn and Khain 2007). The more sophisticated microphysical schemes do a better job of showing fall speed sorting. Yet, improvements in skill scores from changes in physical parameterizations tend to be incremental. As discussed earlier, it seems no particular parameterization or model always provides the best skill.

Another improvement in modeling has been the increasing use of fine grids that do not require application of convective parameterizations. Weisman et al. (2008) have discussed the successes that convection-allowing model runs have had in recent years in predicting convective system mode and diurnal precipitation cycle relative to coarser operational models that were being run at those times. Clark et al. (2009) also demonstrated that convection-allowing simulations have an improved diurnal precipitation cycle relative to simulations that use convective parameterizations. It is believed the better forecasts are directly related to better depiction of convective processes such as upscale growth and propagation mechanisms in the fine grid simulations. Nonetheless, significant errors are still present spatially in most of these simulations, even when morphological evolution and timing are correctly captured. With these spatial errors, objective verification scores remain low.

Because of the general lack of skill for warm season QPF and the large sensitivity of a single deterministic forecast to physical parameterizations, an increasing body of research has been investigating the use of ensemble approaches for precipitation forecasting (e.g. Mullen et al. 1999, Wandishin et al. 2001, Ebert 2001). Ensemble forecasting approaches have long been used for medium range forecasting (typically 3-5 days or more), where the ensembles have usually been created through methods that perturb initial conditions (e.g., Tracton and Kalnay 1993). For these medium range forecasts, errors in the initial conditions make much larger relative contributions than model errors after larger-scale error growth becomes non-linear, roughly after 24 hours (Gilmour 2001). Such an ensemble will reveal useful information about the most likely state of the atmosphere to occur and the uncertainty in the forecast if initialization and observation error dominate over model errors. Probability forecasts are usually easy to generate from such ensembles. In addition, ensemble mean forecasts are usually more skillful than those from any single member (Leith 1974).

However, for short-range warm-season rainfall prediction, it appears model errors may be much larger than other errors. Alhamed et al. (2002), for instance, found that in a mixed model, mixed initial condition ensemble, the members clustered strongly by model (Figure. 5). Gallus and Segal (2001) found that modifications in the initial conditions to better represent mesoscale features did not systematically improve warm season convective system rainfall forecasts in the Upper Midwest, and spread among the runs with the adjusted initial conditions was much less than that occurring when the convective scheme was varied. This result agrees with Stensrud et al. (1999a), among others, who found that perturbed initial conditions may result in limited dispersion in short-range, mesoscale model ensembles. Some studies have found little correlation between ensemble spread and forecast skill for short-range forecasts (Hamill and Colucci 1998, Stensrud et al. 1999a, Hou et al. 2001).

Although it appears forecast skill of ensembles for parameters like warm season precipitation may be increased through the construction of a mixed physics - mixed model ensemble that increases dispersion (Stensrud et al. 2000), questions remain about the interpretation of results from such ensembles. Biases in the models may skew the ensemble prediction, and it is difficult to know if the dispersion present is an accurate reflection of the uncertainty in the forecast.

One possible advantage of using an ensemble approach for warm season convective system rainfall forecasting is that the actual event should end up somewhere within the umbrella of model solutions if the ensemble is constructed well, making an operational forecaster aware of the possibility, albeit small, for the event to occur. As shown in Jankov and Gallus (2004), it is not uncommon to have warm season events where one or two model configurations exhibit almost no skill. Not only are objective measures of skill poor (which might reflect problems applying traditional grid point to grid point verification methods to mesoscale model output, see Mass et al. 2002), these forecasts, subjectively, look nothing like what actually occurs. Reliance upon any deterministic forecast would be foolish in such events. However, an important question becomes, will an ensemble always improve forecastability of the most challenging events? Is it possible that no existing model, observational dataset for initialization, or configuration of physical parameterizations may be able to accurately simulate some events?

One good example is the damaging derecho event discussed briefly above, accompanied by substantial rainfall that occurred after 12 UTC on 4 June 1999 in the Midwestern United States (Gallus et al. 2005). Despite the long-lived nature of the event, and the fact that the initial system was present at 12 UTC, numerical model simulations with both 10 km and 4 km grid spacing did an exceptionally poor job of depicting this convective system. Simulations were performed using the Eta, WRF and MM5 models with a range of physical parameterizations, and three different sources of initial and boundary condition data (from the NCEP Eta, AVN, and RUC models). In addition, mesoscale adjustments were made in the initialization of some runs to better represent mesoscale features and help initiate convection where radar echo was present at initialization time. Although the study emphasized runs initialized at 12 UTC, some additional runs were performed using 06 UTC AVN model output for initialization and boundary condition information. Despite all of the varied models, initial and boundary conditions, and parameterizations used, no simulation indicated the derecho event, suggesting that model error (errors related to model design and not initial and boundary condition data) alone was likely not the primary cause of the poor forecasts. Instead the inability of models to simulate the event implies a much better observation network than what now exists operationally might be needed, or better methods of including additional information from radar and satellite.

It also appears ensembles may need to use convection-allowing grid spacings to adequately capture forcing mechanisms important in generating convective systems. Clark et al. (2010) found that a 4 km grid spacing convection-allowing ensemble of 10 members did a much better job at showing convective redevelopment associated with a mesoscale convective vortex (MCV) than did a 30 member ensemble that used relatively coarse 20 km grid spacing (Figure 6). The 20 km grid ensembles did a poor job of simulating the path taken by the MCV, with large improvement when 4 km grid spacing was used. The computational cost of a convection-allowing ensemble has been prohibitive until recently, but increasing computer power should allow the use of such ensembles operationally within a few years. In fact, a 50 member convection-allowing ensemble was run quasi-operationally by the Center for the Analysis and Prediction of Storms in support of the 2011 National Oceanic and Atmospheric Administration Hazardous Weather Testbed Spring Experiment (S. Weiss, Storm Prediction Center, 2011, personal communication). Recent projects similar to this one (e.g. Schwartz et al. 2010) are investigating ways to best use such high resolution ensemble guidance in the forecasting of weather parameters such as rainfall. These methods of using ensemble information to create both quantitative and probabilistic rainfall forecasts can be thought of as post-processing techniques and are discussed in the next section.

Post-Processing Techniques

Because changes in numerical models themselves generally result in only small improvements in objective skill scores for precipitation forecasts, post-processing is often performed to improve the accuracy or quality of a forecast. Statistical packages like Model Output Statistics (MOS) used by the National Weather Service in the United States have been shown to improve upon forecasts taken directly from the numerical model output (Glahn and Lowry 1972, Bermowitz 1975, Charba and Samplatsky 2011). Although MOS was originally developed for a relatively small number of sites around the country, the newer High Resolution MOS product (Charba and Samplatsky 2001) provides forecast information on a 4 km grid. MOS produces both probability of precipitation (POP) forecasts and most likely category or binned precipitation amount.

Hydrologists and wind energy forecasters have often used statistical techniques (e.g., Burnash et al. 1973) to create forecasts for stream flow and turbine height winds, respectively. Other statistical techniques such as neural networks have been used in some regions, creating more skillful precipitation forecasts than simpler methods such as linear regression models (e.g., Ramirez et al. 2005). Theis (2005) and Schaffer et al. (2011) showed that the most skillful probability forecasts for rainfall may come from techniques that make use of a neighborhood of points in close proximity spatially and/or temporally. As will be seen in the next section, neighborhood approaches are also being investigated as a means of providing more appropriate verification information for fine grid-scale forecasts.

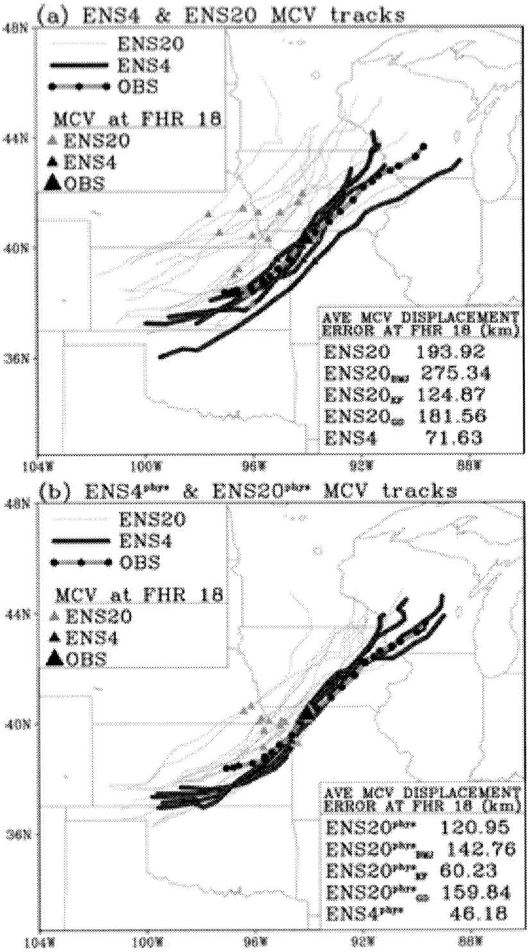

Figure 6. Manually identified MCV tracks for (a) ENS20 (thin gray), ENS4 (thick black), and the RUC analysis (gray line with black dots marked OBS) and (b) ENS20phys (thin gray), ENS4phys (thick black), and the RUC analysis (gray line with black dots marked OBS). Triangles mark the locations of MCVs at forecast hour 18, and average displacement errors for various ensemble subsets at forecast hour 18 are shown in the bottom right of (a) and (b). From Clark et al. (2010), © American Meteorological Society. Reprinted with permission.

Regarding ensembles, although simple arithmetic means may produce good forecasts for some atmospheric variables such as 500 mb height, they tend to smooth precipitation fields too much, resulting in low biases for heavy thresholds and high biases for light ones. Ebert (2001) discusses one method to reduce these problems, the probability matched (PM) ensemble mean. The PM mean is constructed by replacing the PDF of the ensemble mean with the PDF of the *n* ensemble member QPFs, calculated by pooling all of the forecasted precipitation amounts, sorting from largest to smallest, and choosing every *n*th value. Thus, the PM mean assumes the best spatial representation of rainfall is given by the ensemble mean and the best frequency distribution of amounts is given by the individual member QPFs. Because individual ensemble members often overpredict or underpredict rainfall intensity or coverage, bias correction or reduction (e.g., Ebert 2001) may also be performed to improve the quality of ensemble forecasts.

Another advantage of the use of ensembles for precipitation forecasting is that they facilitate the creation of POP forecasts. The simplest approach to generating POPs is to assign equal weighting to each ensemble member (Hamill and Whitaker 2006). Calibration using observed data, however, can improve the POPS. Hamill and Colucci (1997) used a Gumbel distribution to fit the ensemble data while Hamill and Whitaker (2006) calibrated using reforecasts. POPs have also been determined based on the binning of precipitation amounts from forecasts, and this technique can be done using either ensembles (Yussouf and Stensrud 2008, Schaffer et al. 2011) or single deterministic forecasts (e.g. Gallus and Segal 2004).

New Verification Methods

The evaluation of the quality of a forecast is itself a difficult issue, particularly for mesoscale features that dominate precipitation fields during the warm season. Traditionally evaluations have used objective skill measures, often using contingency table inputs for a point to point comparison. It is important to note, however, that no single objective measure of forecast skill can provide complete insight into forecast accuracy (Murphy 1995). Operationally, the most common tools in recent years for this purpose have been two objective measures, bias and Equitable Threat Score (ETS). Although each measure provides some useful information ETSs can be misleading since higher ETSs are often associated with overly high biases (Hamill, 1999).

Mass et al. (2002) discuss how traditional measures may double penalize some fine-grid simulations that are capable of predicting the detailed structure and very intense rainfall that occurs in warm season events but are likely to suffer from a displacement error. If one considers Mean Absolute Error, for instance, a small displacement error for an intense thunderstorm such that none of the forecasted rainfall region overlaps with that observed will earn twice the error of a forecast that was incapable of simulating any rainfall. Error accrues in the former in both the region where precipitation was observed and not forecasted, and the region where it was forecasted and not observed, whereas in the latter, the only error is from observed precipitation that was not forecasted. An example showing how Critical Success Index (CSI) might yield a different impression than that determined subjectively is shown in Figure 7 (from Davis et al. 2006). The highest CSI is found only in the forecast where some rain overlaps the area where it was observed, even though the shape of the rain area is very different from that observed, and the bias is as bad as in any of the cases shown. Simulations depicting the proper shape of the rain area and having a perfect bias end up with CSI values of zero. Traditional verification scores provide incomplete information about the quality of a forecast because they only make comparisons on a point-to-point basis with no regard to spatial information (Baldwin and Kain 2006, Casati et al. 2008).

To reduce some of the problems associated with traditional verification measures and produce quantitative verification measures that might better agree with subjective impressions, several different approaches have been taken. A comprehensive review of these can be found in Gilleland et al. (2009) who describes the Intercomparison of Spatial Forecast Verification Methods Project, and this chapter will only briefly mention some of the approaches.

New approaches for verification can mostly be classified into two overall categories: filtering methods and displacement methods (Gilleland et al. 2009). The filtering methods can be further delineated into neighborhood and scale separation, and the displacement methods can be divided into features based and field deformation. The filtering methods both apply a spatial filter to one or both fields (or sometimes to the difference field), and then calculate verification statistics on the filtered fields. The filter is usually applied at progressively coarser scales to provide information about the scale(s) at which the forecast has skill. The neighborhood methods apply a smoothing filter, whereas the scale-separation techniques apply several single-bandpass spatial filters (Fourier, wavelets, etc.) so that performance at separate scales can be evaluated independently.

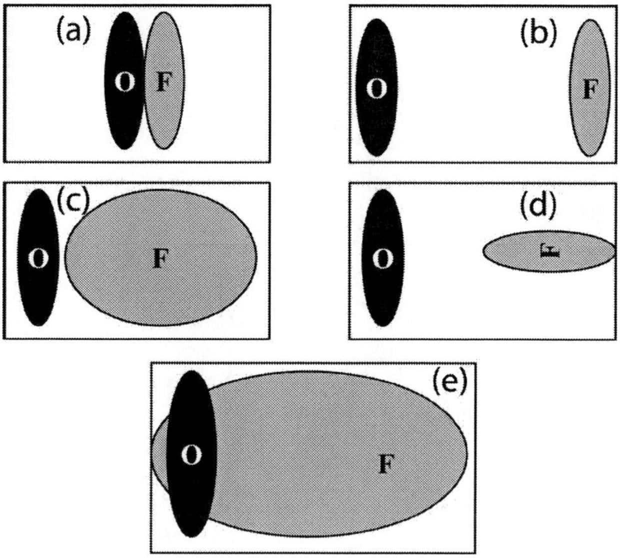

Figure 7. A schematic example of various forecast and observation combinations. (a)–(d) These all yield CSI = 0, whereas (e) has positive CSI, but would probably not be evaluated as the best subjectively. From Davis et al. (2006), © American Meteorological Society. Reprinted with permission.

The features-based and field deformation categories are similar in that they both try to "fit" the forecast to the observations as well as possible (Gilleland et al. 2009), give information about how much the forecast field needs to be manipulated spatially (displacement, rotations, scaling, etc.) and quantify the residual errors to obtain a more meaningful notion of skill. One increasingly commonly used features-based verification package is the Method for Object-based Diagnostic Evaluation (MODE; Davis et al. 2006). An example of MODE output for one 6 hour forecast period of accumulated rainfall is shown in Figure 8. This type of verification can provide users with more information on discrepancies between forecasts and observations than is typically available from traditional objective measures such as ETS or bias, but is often criticized because results are often sensitive to numerous user-defined parameters that influence the selection and matching of objects. Although most of these newer techniques were designed with verification of single deterministic forecasts in mind, preliminary use of object-based methods to evaluate ensembles has begun (Gallus 2010). In that study, the object parameters of rain rate, areal coverage, rain volume, and displacement errors were compared among two different ensembles that had been studied using more traditional techniques (Clark et al. 2008, 2009).

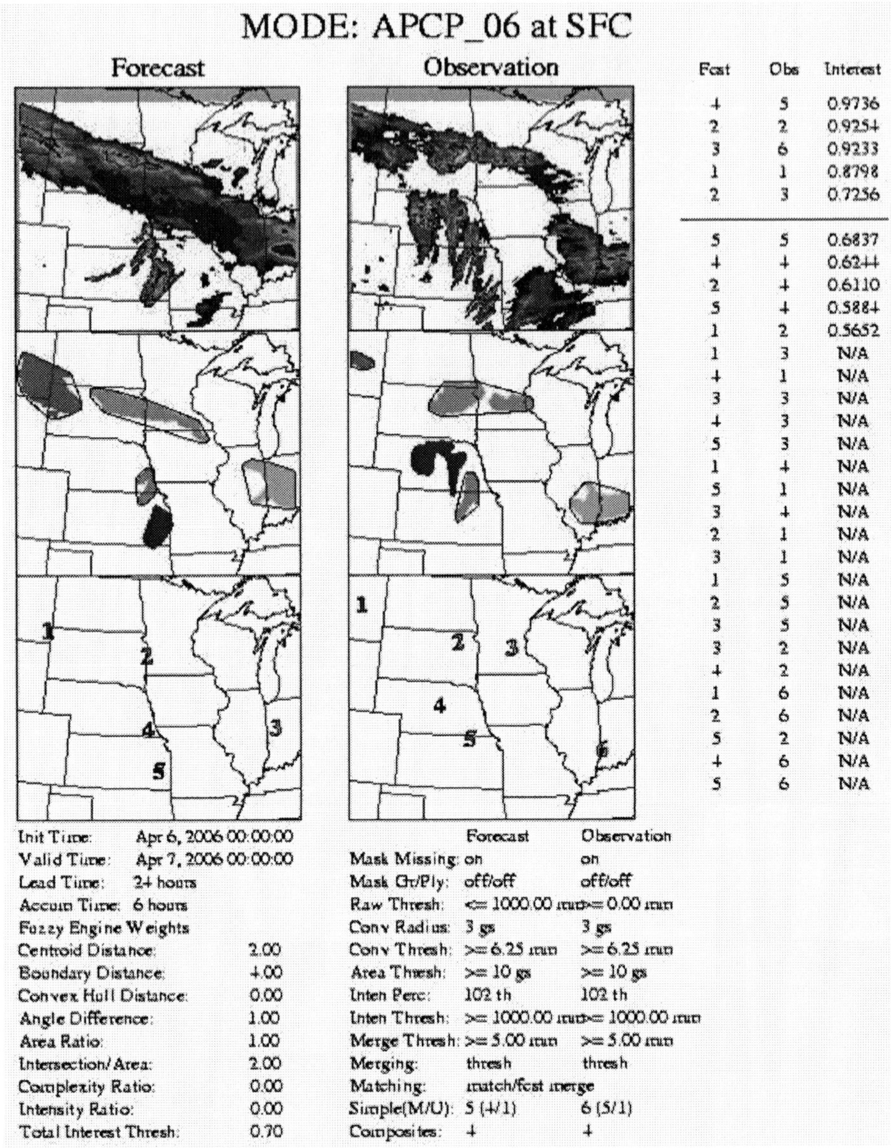

Figure 8. Example of MODE output for 6 h precipitation accumulation during 18-00 UTC April 6-7, 2006. Middle column demonstrates objects identified using 6.25 mm threshold with darkest colors representing missed (object 4 in observations, see bottom right panel) or false alarm (object 5 in forecast in bottom left panel) systems. Other systems were matched. Interest value (measure of similarity) for comparisons of various systems is shown in data at right.

Conclusion

Although substantial improvement has been made in general in the prediction of precipitation in the last few decades, warm season precipitation has remained a difficult challenge. Skill scores are much lower for these forecasts than their cold season counterparts, and improvement has been slower as well. The small-scale of most warm season precipitation-producing systems along with the small-scale of the atmospheric phenomena that typically trigger these events are primarily to blame. Although increasing computer power has allowed numerical weather prediction models to run with increasingly fine grid spacing, parameterizations of processes important in the generation of precipitation still introduce errors into the forecasts that can be large.

Efforts to improve warm season precipitation forecasting have focused on improved initialization, modeling techniques, and post-processing approaches. The assimilation of radar information appears to offer promising improvements in forecast skill, although the improvements are generally limited to the first 12 hours of a forecast and may not be pronounced in situations where there is little echo coverage at initialization time and convective development occurs afterwards. Because rainfall forecasts are very sensitive to moisture content, new observing systems that better depict small-scale variations in moisture may offer more hope for future improvements in skill. The use of ensembles may also improve precipitation forecasting, although questions remain about the best design for these ensembles. Although the ideal approach to improve warm season precipitation forecasting is to improve the initialization of and the dynamic cores and parameterizations within the models themselves, some gain in skill can be achieved also through post-processing techniques that act to reduce errors, such as adjustment for biases in the model output.

Finally, the fine scale variations occurring spatially and temporally in convective precipitation make traditional objective verification measures less useful for warm season precipitation verification than for larger-scale cold season events. To better determine the skill of warm season precipitation forecasts, new verification techniques are being used. An added advantage of these techniques is that the extra information they provide may prove valuable to model developers, helping to pinpoint specific shortcomings in the models to potentially accelerate needed improvements in precipitation forecasting skill.

REFERENCES

Alhamed, A., S. Lakshmivarahan, and D. J. Stensrud, 2002: Cluster analysis of multimodel ensemble data from SAMEX. *Mon. Wea. Rev.*, 130, 226–256.

Atkins, N. T., R. M. Wakimoto, and C. L. Ziegler, 1998: Observations of the Finescale Structure of a Dryline during VORTEX 95. *Mon. Wea. Rev.*, 126, 525-550.

Baik, J., Y. Kim, and H. Chun, 2001: Dry and Moist Convection Forced by an Urban Heat Island. *J. App. Meteor.*, 40, 1462-1475.

Baldwin, M. E. and J. S. Kain, 2006: Sensitivity of several performance measures to displacement error, bias, and event frequency. *Wea. Forecasting*, 21, 636-648.

Bermowitz, R. J., 1975: An application of model output statistics to forecasting quantitative precipitation. *Mon. Wea. Rev.*, 103, 149-153.

Bryan, G. H., J. C. Wyngaard, and J. M. Fritsch, 2003: Resolution requirements for the simulation of deep moist convection. *Mon. Wea. Rev.*, 131, 2394-2416.

Bukovsky, M. S., J. S. Kain, and M. E. Baldwin, 2006: Bowing convective systems in a popular operational model: Are they for real? *Wea. Forecasting*, 21, 307-324.

Burnash, R. J. C., R. L. Ferral, and R. A. McGuire, 1973: *A generalized streamflow simulation system: Conceptual models for digital computers,* National Weather Service, NOAA, and the State of California Department of Water Resources technical report: Joint Federal-State River Forecast Center.

Byers, H. R., and R. R. Braham Jr., 1949: The Thunderstorm. U.S. Government Printing Office, 287 pp.

Casati, B., L. J. Wilson, D. B. Stephenson, P. Nurmi, A. Ghelli, A. Pocernich, M. Damrath, E. E. Ebert, B. G. Brown, and S. Mason, 2008: Forecast verification: current status and future directions. *Met. Apps.*, 15, 3-18.

Charba, J. P., and F. G. Samplatsky, 2011: High-resolution GFS-based MOS quantitative precipitation forecasts on a 4-km grid. *Mon. Wea. Rev.*, 139, 39-68.

Clark, A. J., W. A. Gallus Jr., and T-C. Chen, 2008: Contributions of mixed physics versus perturbed initial/lateral boundary conditions to ensemble-based precipitation forecast skill. *Mon. Wea. Rev.*, 136, 2140–2156.

Clark, A. J., W. A. Gallus Jr., and M. Xue, 2009: A comparison of precipitation forecast skill between small convection-resolving and large non-convection-resolving ensembles. *Wea. Forecasting*, 24, 1121–1140.

Clark, A. J., W. A. Gallus, Jr., M. Xue, and F. Kong, 2010: Convection-allowing and convection-parameterizing ensemble forecasts of a mesoscale convective vortex and associated severe weather. *Wea. Forecasting*, 25, 1052-1081.

Clark, C. A., and R. W. Arritt, 1995: Numerical Simulations of the Effect of Soil Moisture and Vegetation Cover on the Development of Deep Convection. *J. Applied Meteor.*, 34, 2029-2045.

Colle, B. A., and C. F. Mass, 2000: The 5-9 February 1996 flooding event over the Pacific Northwest: Sensitivity studies and evaluation of the MM5 precipitation forecasts. Mon. Wea. Rev., 128, 593-617.

Colman, B. R., 1990a: Thunderstorms above Frontal Surfaces in Environments without Positive CAPE. Part I: Climatology. *Mon. Wea. Rev.*, 118, 1103-1122.

Colman, B. R., 1990b: Thunderstorms above Frontal Surfaces in Environments without Positive CAPE. Part II: Organization and Instability Mechanisms. *Mon. Wea. Rev.*, 118, 1123-1144..

Cotton, W. R., M.-S. Lin, R. L. McAnelly, and C. J. Tremback, 1989: A composite model of mesoscale convective complexes. *Mon. Wea. Rev.*, 117, 765-783.

Crook, N. A., 1996: Sensitivity of Moist Convection Forced by Boundary Layer Processes to Low-Level Thermodynamic Fields. *Mon. Wea. Rev.*, 124, 1767-1785.

Crook, N. A., and J. B. Klemp, 2000: Lifting by convergence lines. J. Atmos. Sci, 57, 873–890.

Davis, C. A., B. Brown, and R. Bullock, 2006: Object-based verification of precipitation forecasts. Part I: Methodology and application to mesoscale rain areas. Mon. Wea. Rev., 134, 1772-1784.

Doswell, C. A., III, H. E. Brooks, and R. A. Maddox, 1996: Flash flood forecasting: An ingredients-based methodology. *Wea. Forecasting*, 11, 560-581.

Ebert, E. E., 2001: Ability of a poor man's ensemble to predict the probability and distribution of precipitation. *Mon. Wea. Rev.*, 129, 2461-2480.

Faller, A. J., 1965: Large Eddies in the Atmospheric Boundary Layer and Their Possible Role in the Formation of Cloud Rows. *J. Atmos. Sci.*, 22, 176-184.

Ferrare, R. A., J. L. Schols, E. W. Eloranta, and R. Coulter, 1991: Lidar Observations of Banded Convection during BLX83. *J. Applied. Meteor.*, 30, 312-326.

Fritsch, J. M., R. J. Kane, and C. R. Chelius, 1986: The contribution of mesoscale convective weather systems to the warm-season precipitation in the United States. *J. Climate Appl. Meteor.*, 25, 1333-1345.

Gallus, W. A., Jr., 1999: Eta simulations of three extreme precipitation events: Impact of resolution and choice of convective parameterization. *Wea. Forecasting*, 14, 405-426.

Gallus, W. A., Jr., 2010: Application of Object-Based Verification Techniques to Ensemble Precipitation Forecasts. Wea. Forecasting, 25, 144–158. doi: 10.1175/2009WAF2222274.1

Gallus, W. A. and J. F. Bresch, 1997: An intense small-scale wintertime vortex in the midwest United States. Mon. Wea. Rev., 125, 2787-2807.

Gallus, W. A., Jr., and R. H. Johnson, 1991: Heat and moisture budgets budgets of an intense midlatitude squall line. *J. Atmos. Sci.*, 48, 122-146.

Gallus, W. A., Jr., and M. Segal, 2000: Sensitivity of Forecast Rainfall in a Texas Convective System to Soil Moisture and Convective Parameterization. *Wea. Forecasting*, 15, 509-525.

Gallus Jr., W. A., and M. Segal, 2001: Impact of improved initialization of mesoscale features on convective system rainfall in 10-km Eta simulations. *Wea. Forecasting*, 16, 680–696.

Gallus, W. A., Jr., and M. Segal, 2004: Does increased predicted warm-season rainfall indicate enhanced likelihood of rain occurrence? *Wea. Forecasting*, 19, 1137-1135.

Gallus Jr., W. A., J. C. Correia, and I. Jankov, 2005: The 4 June 1999 derecho event: A particularly difficult challenge for numerical weather prediction. Wea. Forecasting, 20, 705–728.

Gilleland, E., D. Ahijevych, B. G. Brown, B. Casati, and E. E. Ebert, 2009: Intercomparison of Spatial Forecast Verification Methods. *Wea. Forecasting*, 24, 1416–1430. doi: 10.1175/2009WAF2222269.1

Gilmore, M. S., J. M. Straka, and E. N. Rasmussen, 2004: Precipitation and evolution sensitivity in simulated deep convective storms: Comparisons between liquid-only and simple ice and liquid phase microphysics. *Mon. Wea. Rev.*, 132, 1897-1916.

Gilmour, I., L. A. Smith, and R. Buizza, 2001: Linear region duration: Is 24 hours a long time in synoptic weather forecasting? *J. Atmos. Sci.*, 58, 3525-3539.

Glahn, H. R., and D. A. Lowry, 1972: The use of model output statistics (MOS) in objective weather forecasting. *J. Appl. Meteor.*, 11, 1203-1211.

Hamill, T. M., 1999: Hypothesis test for evaluating numerical precipitation forecasts. Wea. Forecasting, 14, 155–167.

Hamill, T. M., and S. J. Colucci, 1997: Verification of Eta-RSM short-range ensemble forecasts. *Mon. Wea. Rev.*, 125, 1312-1327.

Hamill, T. M., and S. J. Colucci, 1998: Evaluation of Eta/RSM Ensemble Probabilistic Precipitation Forecasts. *Mon. Wea. Rev.*, 126, 711-724.

Hamill, T. M., and J. S. Whitaker, 2006: Probabilistic quantitative precipitation forecasts based on reforecast analogs: Theory and application. Mon. Wea. Rev., 134, 3209-3229.

Hill, G. E., 1977: Initiation Mechanisms and Development of Cumulus Convection. *J. Atmos. Sci.*, 34, 1934-1941

Hou, D., E. Kalnay, and K. K. Droegemeier, 2001: Objective verification of the SAMEX '98 ensemble forecasts. Mon. Wea. Rev., 129, 73-91.

Houze, R. A., Jr., 2004: Mesoscale convective systems. *Rev. Geophys.*, 42, RG4003, doi:10.1029/2004RG000150.

Jankov, I,. and W. A. Gallus, Jr., 2004: Contrast between good and bad forecasts of warm season MCS rainfall. *J. Hydrology*, 288, 122-152.

Jankov, I., W. A. Gallus, Jr., B. Shaw, and S. E. Koch, 2005: On the impacts of different WRF physical parameterizations and their interactions on warm season MCS rainfall. *Wea. Forecasting*, 20, 1048-1060.

Johnson, R. H., and P. J. Hamilton, 1988: The relationship of surface pressure features to the precipitation and air flow structure of an intense midlatitude squall line. *Mon. Wea. Rev.*, 116, 1444-1472.

Kain, J. S., and J. M. Fritsch, 1992: The role of the convective "trigger function" in numerical forecasts of mesoscale convective systems. *Meteor. Atmos. Phys.*, 49, 93-106.

Kain, J. S., and J. M. Fritsch, 1998: Multiscale convective overturning in mesoscale convective systems: Reconciling observations, simulations and theory. *Mon. Wea. Rev.*, 126, 2254-2273.

Kain J. S., S. J. Wicker, J. J. Levit, M. E. Baldwin, and D. R. Bright, 2006: Examination of convection-allowing configurations of the WRF model for the prediction of severe convective weather: The SPC/NSSL Spring Program 2004. *Wea. Forecasting*, 21, 167-181.

Kuo, Y.-H., R. J. Reed, and Y. Liu, 1996: The ERICA IOP5 storm: Mesoscale cyclogenesis and precipitation parameterization. *Mon. Wea. Rev.*, 124, 1409-1434.

Lean, H. W., P. A. Clark, M. Dixon, N. M. Roberts, A. Fitch, R. Forbes, and C. Halliwell, 2008: Characteristics of high resolution versions of the MET Office Unified Model for forecasting convection over the UK. *Mon. Wea. Rev.* 136, 3408-3424.

Leith, C. E., 1974: Theoretical skill of Monte Carlo forecasts. *Mon. Wea. Rev.*, 102, 409-418.

LeMone, M. A., and R. J., Meitin, 1984: Three Examples of Fair-Weather Mesoscale Boundary-Layer Convection in the Tropics. *Mon. Wea. Rev.*, 112, 1985-1998.

Lilly, D. K., 1966: On the Instability of Ekman Boundary Flow. *J. Atmos. Sci.*, 23, 481-494.

Lynn, B., and A. Khain, 2007: Utilization of spectral bin microphysics and bulk parameterization schemes to simulate the cloud structure and precipitation in a mesoscale rain events. *J. Geophys. Res.*, 112, D22205, doi:10.1029/2007JD008475.

Maddox, R. A., 1983: Large-scale meteorological conditions associated with mid-latitude, mesoscale convective complexes. *Mon. Wea. Rev.*, 111, 1475-1493.

Mass, C.F., D.Ovens, K. Westrick and B.A. Colle, 2002: Does increasing horizontal resolution produce more skillful forecasts? *Bull. Amer. Meteor. Soc.*, March 2002, 407-430.

Milbrandt, J. A., and M. K. Yau, 2005: A multimoment bulk microphysics parameterization. Part I: Analysis of the role of the spectral shape parameter. *J. Atmos. Sci.*, 62, 3051-3064.

Molinari, J. and M. Dudek, 1986: Implicit versus explicit convective heating in numerical weather prediction models. *Mon. Wea. Rev.*, 114, 1822-1831.

Molinari, J. and M. Dudek, 1992: Parameterization of convective precipitation in mesoscale numerical models: A critical review. *Mon. Wea. Rev.*, 120, 326-344.

Moore, J. T., F. H. Glass, C. E. Graves, S. M. Rochette, and M. J. Singer, 2003: The environment of warm-season elevated thunderstorms associated with heavy rainfall over the central United States. *Wea. Forecasting*, 18, 861–878.

Morrison, H. G., G. Thompson, and V. Tatarskii, 2009: Impact of cloud microphysics on the development of trailing stratiform precipitation in a simulated squall line: Comparison of one- and two- moment schemes. *Mon. Wea. Rev.*, 137, 991-1007.

Mullen, S. L., J. Du, and F. Sanders, 1999: The dependence of ensemble dispersion on analysis–forecast systems: Implication to shortrange ensemble forecasting of precipitation. *Mon. Wea. Rev.*, 127, 1674–1686.

Murphy, A.H., 1995: A coherent method of stratification within the general framework for forecast verification. *Mon. Wea. Rev.*, 123, 1582-1588.

Nakanishi, M., and H. Niino, 2004: An improved Mellor-Yamada level-3 model: Its numerical stability and application to a regional prediction of advection fog. *Bound.-Layer Meteor.*,119, 397-407.

Olson, D. A., N. W. Junker, and B. Korty, 1995: Evaluation of 33 years of quantitative precipitation forecasting at the NMC. *Wea. Forecasting*, 10, 498-511.

Potter, B. E., 1991: Improvements to a commonly used cloud microphysical bulk parameterization, *J. Appl. Meteor.*, 30, 1040-1042.

Purdom, J. F. W., 1976: Some Uses of High-Resolution GOES Imagery in the Mesoscale Forecasting of Convection and Its Behavior. *Mon. Wea. Rev.*, 104, 1474-1483.

Ramirez, M. C. V., H. F. de Campo Velho, and N. J. Ferreira, 2005: Artificial neural network technique for rainfall forecasting applied to the Sao Paulo region. *J. of Hydrology*, 301, 1-4, 146-162.

Rogers, R. F., J. M. Fritsch, and W. C. Lambert, 2000: A simple technique for using radar data in the dynamic initialization of a mesoscale model. *Mon. Wea. Rev.*, 128, 2560-2574.

Schaefer, J. T., 1986: Severe Thunderstorm Forecasting: A Historical Perspective. *Wea. Forecasting*, 1, 164-189.

Schaffer, C. J., W. A. Gallus, Jr., and M. Segal, 2011: Improving probabilistic ensemble forecasts of convection through the application of QPF-POP relationships. *Wea. Forecasting*, 26, (in press).

Schwartz, C. S., J. S. Kain, S. J. Weiss, M. Xue, D. R. Bright, F. Kong, K. W.Thomas, J. J. Levit, M. C. Coniglio, and M. S. Wandishin, 2010: Toward improved convection-allowing ensembles: model physics sensitivities and optimizing probabilistic guidance with small ensemble membership. *Wea. Forcasting*, 25, 263-280.

Segal, M., and R. W. Arritt, 1992: Nonclassical Mesoscale Circulations Caused by Surface Sensible Heat-Flux Gradients. *Bull. Amer. Meteor. Soc.*, 73, 1593-1604.

Shapiro, M. A., 1981: Frontogenesis and Geostrophically Forced Secondary Circulations in the Vicinity of Jet Stream-Frontal Zone Systems. *J. Atmos. Sci.*, 38, 954-973.

Shapiro, M. A., 1982: Mesoscale weather systems of the central United States, CIRES/NOAA Report, Boulder, CO, 78 pp.

Spencer, P. L., and D. J. Stensrud, 1998: Simulating flash flood events: Importance of the sub-grid representation of convection. *Mon. Wea. Rev.*, 126, 2884-2912.

Stensrud, D. J., 1996: Importance of low-level jets to climate: A review. *J. Climate*, 9, 1698-1711.

Stensrud, D. J., 2007: Parameterization schemes: Keys to understanding numerical weather prediction models. Cambridge University Press, 488 pp.

Stensrud, D. J., and R. A. Maddox, 1988: Opposing Mesoscale Circulations: A Case Study. Wea. Forecasting, 3, 189-204.

Stensrud, D. J., and J. M. Fritsch, 1994: Mesoscale convective systems in weakly forced large-scale environments. Part III: Numerical simulations and implications for operational forecasting. *Mon. Wea. Rev.*, 122, 2084-2104.

Stensrud, D.J., H.E. Brooks, J. Du, M.S. Tracton, and E. Rogers, 1999a: Using ensembles for short-range forecasting. *Mon. Wea. Rev.*, 127, 433-446.

Stensrud, D.J., G.Manikin, E. Rogers, and K. Mitchell, 1999b: Importance of cold pools to mesoscale model forecasts. *Wea. Forecasting*, 14, 650-670.

Stensrud, D. J., J-W. Bao, and T. T. Warner, 2000: Using initial condition and model physics perturbations in short-range ensemble simulations of mesoscale convective systems. Mon. Wea. Rev., 128, 2077–2107.

Sun, J., and N. A. Crook, 1997: Dynamical and microphysical retrieval from Doppler radar observations using a cloud model and its adjoint. Part I: Model development and simulated data experiments. *J. Atmos. Sci.*, 54, 1642-1661.

Takano, I., and A. Segami, 1993: Assimilation and initialization of a mesoscale model for improved spin-up of precipitation. *J. Meteor. Soc. Japan*, 71, 377-391.

Theis, S. E., A. Hense, and U. Damrath, 2005: Probabilistic precipitation forecasts from a deterministic model: a pragmatic approach. *Meteorological Applications*, 12, 257-268.

Tong, M., and M. Xue, 2005: Ensemble Kalman filter assimilation of Doppler radar data with a compressible nonhydrostatic model: OSS experiments. *Mon. Wea. Rev.*, 133, 1789-1807.

Tracton, M. S., and E. Kalnay, 1993: Ensemble prediction at the National Meteorological Center: Practical aspects. *Wea. Forecasting*, 8, 379-398.

Trier, S. B., D. B. Parsons, and J. H. Clark, 1991: Environment and Evolution of a Cold-Frontal Mesoscale Convective System. *Mon. Wea. Rev.*, 119, 2429-2455.

Wandishin, M. S., S. L. Mullen, D. J. Stensrud, and H. E. Brooks, 2001: Evaluation of a short-range multimodel ensemble system. Mon. Wea. Rev., 129, 729–747.

Wang, W. and T. T. Warner, 1988: Use of four-dimensional data assimilation by Newtonian relaxation and latent-heat forcing to improve a mesoscale-model precipitation forecast: A case study. *Mon. Wea. Rev.*, 116, 2593-2613.

Wang, W., and N. L. Seaman, 1997: A comparison study of convective parameterization schemes in mesoscale model. *Mon. Wea. Rev.*, 125, 252-278.

Weckwerth, T. M., J. W. Wilson, and R. M. Wakimoto, 1996: Thermodynamic variability within the convective boundary layer due to horizontal convective rolls. *Mon. Wea. Rev.*, 124, 769-784.

Weckwerth, T. M., J. W. Wilson, R.M. Wakimoto, and N. A. Crook, 1997: Horizontal convective rolls: Determining the environmental conditions supporting their existence and characteristics. *Mon. Wea. Rev.*, 125, 505–526.

Weckwerth, T. M., H. V. Murphey, C. Flamant, J. Goldstein, C. R. Pettet, 2008: An Observational Study of Convection Initiation on 12 June 2002 during IHOP_2002. *Mon Wea. Rev.*, 136, 2283-2304.

Weisman, M. L., W. C. Skamarock, and J. B. Klemp, 1997: The resolution dependence of explicitly modeled convective systems. *Mon Wea. Rev.*, 125, 527-548.

Weisman, M. L., C. Davis, W. Wang, K. W. Manning, and J. B. Klemp, 2008: Experiences with 0–36-h Explicit Convective Forecasts with the WRF-ARW Model. *Wea. Forecasting*, 23, 407–437. doi: 10.1175/2007WAF2007005.1

Wilson, J. W., and R. D. Roberts, 2006: Summary of convective storm initiation and evolution during IHOP: Observational and modeling perspective. *Mon. Wea. Rev.*, 134, 23-47.

Wilson, J. W., N. A. Crook, C. K. Mueller, J. Sun, and M. Dixon, 1998: Nowcasting thunderstorms: A status report. *Bull. Amer. Meteor. Soc.*, 79, 2079–2099.

Xue, M., D.-H. Wang, J.-D. Gao, K. Brewster, and K. K. Droegemeier, 2003: The Advanced Regional Prediction System (ARPS), storm-scale numerical weather prediction and data assimilation. *Meteor. Atmos. Physics*, 82, 139-170.

Xue, M., F. Kong, K. W. Thomas, J. Gao, Y. Wang, K. Brewster, K. K. Droegemeier, X. Wang, J. Kain, S. Weiss, D. Bright, M. Coniglio, and J. Du, 2009: CAPS realtime multi-model convection-allowing ensemble and 1-km convection-resolving forecasts for the NOAA Hazardous Weather Testbed 2009 Spring Experiment. *23nd Conf. Wea. Anal.*

Forecasting/19th Conf. Num. Wea. Pred., Omaha, NB, Amer. Meteor. Soc., Paper 16A.2.

Yussouf, N. and D. J. Stensrud, 2008: Reliable probabilistic quantitative precipitation forecasts from a short-range ensemble forecasting system during the 2005/06 cool season. *Mon. Wea. Rev.*, 136, 2157-2172.

Ziegler, C. L., T. J. Lee, R. A. Pielke, Sr., 1997: Convective Initiation at the Dryline: A Modeling Study. *Mon. Wea. Rev.*, 125, 1001-1026.

Ziegler, C. L. and E. N. Rasmussen, 1998: The initiation of moist convection at the dryline: Forecasting issues from a case study perspective. *Wea. Forecasting*, 13, 1106–1131.

In: Rainfall Forecasting
Editor: T.S.W. Wong, pp. 161-197
ISBN: 978-1-61942-134-9

Chapter 4

THE PRECIPITATION FIELD IN SOUTHERN ITALY: CLIMATOLOGY, PHYSICAL FACTORS AND FORECASTING TECHNIQUES

S. Federico[1,2,*]

[1] Institute of Atmospheric and Climate Sciences – National Research Council (ISAC-CNR), Operative Unit of Support of Lamezia Terme, Lamezia Terme CZ, Italy.

[2] Consorzio per la Ricerca e l'Applicazione di Tecnologie Innovative (CRATI), Rende CS, Italy.

ABSTRACT

This chapter revisits some aspects of the precipitation field, including its prediction, over Calabria peninsula, which is the southwest tip of the Italian boot. Calabria offers an interesting scenario for mesoclimatic studies in spatial and temporal rainfall variability. The region is located in the Central Mediterranean Basin, which has a strong seasonal variability that mirrors in large precipitation differences among seasons. Moreover, the complex orography of the country leads to different exposures of the territory to the rain-bearing air masses and large differences in rainfall can occur in few tens of kilometres.

After presenting the climatology of the rain field over the country and, specifically of heavy rainfall events (precipitation greater than 64

[*] E-mail: s.federico@isac.cnr.it

mm/day), the numerical analysis of an intense and destructive storm that occurred on 10–12 December 2003 was used to study the role of the physical factors involved. This case study was chosen because it is a prototype of heavy rainfall events occurring over the peninsula and more generally in the Mediterranean area.

The factor separation technique was applied to this storm and the case study clearly shows the effect of the three main physical factors involved in intense and destructive rainstorm occurring over Calabria: (1) the orography, (2) the sea, and (3) the synoptic scale forcing.

Results show that the Calabrian orography and the surface latent heat flux were important elements for the precipitation enhancement at the mesoscale. Humid marine air masses were advected by the synoptic flow toward the Calabrian steep coastal mountain ranges determining intense and abundant rainfall.

The analysis of the storm also reveals the crucial role of the synoptic-scale forcing, which can be accounted by considering a single key variable: the potential vorticity (PV). In the extra-tropical atmosphere, intense cyclones often develop under upper-tropospheric jet streams that are characterized by anomalously high values of the PV, and accurate specification of the PV distribution is necessary for the successful prediction of hazardous weather events.

The sensitivity of intense rainstorms to the PV streamer paves the way to a more advanced, yet simple, forecasting technique, based on a Limited Area Model Ensemble Prediction System (LEPS). The technique is particularly useful in data void regions as Calabria, where the presence of the sea makes ground based observations problematic.

A LEPS forecast, whose resolution is able to represent structures of the precipitation field not resolved by global models, generates different scenarios compatible with model error and initial/boundary conditions error. For each scenario, a pseudo water-vapour image is computed by the temperature of the isosteric surface of 75 mg/kg. Pseudo water-vapour images are then compared with the METEOSAT water vapour image, before the time of the expected rainfall in the target area, to classify the scenarios' reliability. The precipitation forecast by these scenarios can be used to refine the deterministic forecast.

The methodology was applied to the 10-12 December 2003 rainstorm. For this case it was able to discard the only scenario predicting comparatively lower rainfall, and to show the good performance of the unperturbed forecast.

Keywords: Calabria peninsula, precipitation climatology, meteorological models, factor separation technique, quantitative precipitation forecasting, Potential Vorticity, Limited area model Ensemble Prediction System, METEOSAT images.

1. INTRODUCTION

The Calabria region (Figure 1) is the southern most tip of the Italian peninsula located in the central Mediterranean Basin, and extends between 37°45' and 40°00' latitude North and between 15°30' and 17°15' longitude East. The region is bounded by the Tyrrhenian Sea (west) and by the Ionian Sea (east and south). The Apennines mountains run north-south along the peninsula and are characterized by five main topographical features reaching 1500–2000 m elevations: Pollino, Catena Costiera, Sila, Serre and Aspromonte. The average width of the region is about 50 km in the west-east direction and 300 km in the north-south direction. There are three main planes by the sea (Sibari, Gioa Tauro, Lamezia) where most of agricultural and industrial areas are found.

Precipitation over mountains is very challenging, especially in case of complex orography (Mass et al., 2003), such as in Calabria. For this peninsula, there are three key physical factors playing a major role in the precipitation: (1) the orography, (2) the presence of the Mediterranean Sea, and (3) the large-scale atmospheric patterns.

Colacino et al. (1997), in a study based on nine raingauges, found that the orography controls the rainfall amount and distribution over the region. They show that the interaction between the orography and mesoscale circulations led to a precipitation gradient between the Tyrrhenian and Ionian side of the peninsula, i.e. precipitation was enhanced in favourably exposed areas (Tyrrhenian side) and suppressed in sheltered areas (Ionian side). In the same study, the important role of the orographic uplift was evident, by comparing the precipitation recorded at a mountainous station over Sila (1800 m a.s.l.) to the precipitation at the stations located near the coast.

Numerical simulations of severe weather have shown the crucial role of the orography, which enhances rainfall in localized spots (Federico et al. 2003a, 2003b) or forces secondary cyclogenesis persisting on the Ionian side of the peninsula (Federico et al., 2007).

The central Mediterranean Sea reaches its maximum temperature during summer and early fall ensuring large water vapour availability for this period of the year. These conditions favour convective instability that can be released through organized convection, if suitable synoptic and mesoscale conditions are met, as in the case of cold intrusion of air masses from the Balkans. These peculiar features of Calabria, i.e. the presence of high mountains nearby a warm sea, can determine favourable conditions for heavy rainfall events.

Precipitation over Calabria has a strong seasonal dependence and large-scale atmospheric patterns play an important role. In summer, the Azores anticyclone extends over the peninsula favouring a rather dry and temperate climate, while breeze circulations develop along the coasts (sea-land breezes) and inland (anabatic and katabatic winds) (Federico et al., 2010). In winter, fall, and spring, Azores anticyclone retreats while the Siberian high extends over North Europe and Scandinavia. This condition favours the development of Mediterranean cyclones, that either enter the Basin from outside or are locally generated, often under the influence of a pre-existing cyclonic system. These cyclones reach Calabria determining a colder and more humid climate compared to summer.

The peculiar characteristics determining the climatology of the precipitation field over Calabria are found in a case study occurred between 10 and 12 December 2003. The storm is studied at the mesoscale by means of the Regional Atmospheric Modeling System (RAMS)[1]. The model was successful at representing the observed accumulated rainfall patterns. This justifies its use for further analysis of the physical and dynamical factors involved. In particular, RAMS is used to evaluate the roles of the Calabrian orography and of the surface latent heat flux through the factor separation technique.

The key role of the synoptic scale forcing and, particularly, of the potential vorticity (PV) streamer, which forced the entire meteorological system from the upper levels, is assessed by a Limited Area Model Ensemble Prediction System (LEPS), which uses the RAMS model to increase the resolution of the European Centre for Medium-Range Weather Forecasts (ECMWF) global model ensemble. Results of the case study show the sensitivity of the rainfall field to the synoptic scale forcing.

The high impact of the synoptic scale forcing on the precipitation field of moderate-heavy rainstorms is the working hypothesis of a simple technique for improving the deterministic forecast. In particular, a LEPS generates five different scenarios by using the height of the dynamical tropopause as the discriminating variable. Pseudo water vapour images of each scenario are compared to the corresponding METEOSAT 7 water vapour image at a specific time, antecedent to the rain occurrence over Calabria, to evaluate the reliability of the different precipitation scenarios. The results show the usefulness of the methodology.

---

[1] RAMS is used since January 2005 to issue weather forecast over Calabria, nowadays at 2.5 km horizontal resolution (http://meteo.crati.it/previsioni.html). This forecast is also used by the Regional Agency for Environmental Protection (ARPACAL) for civil protection purposes.

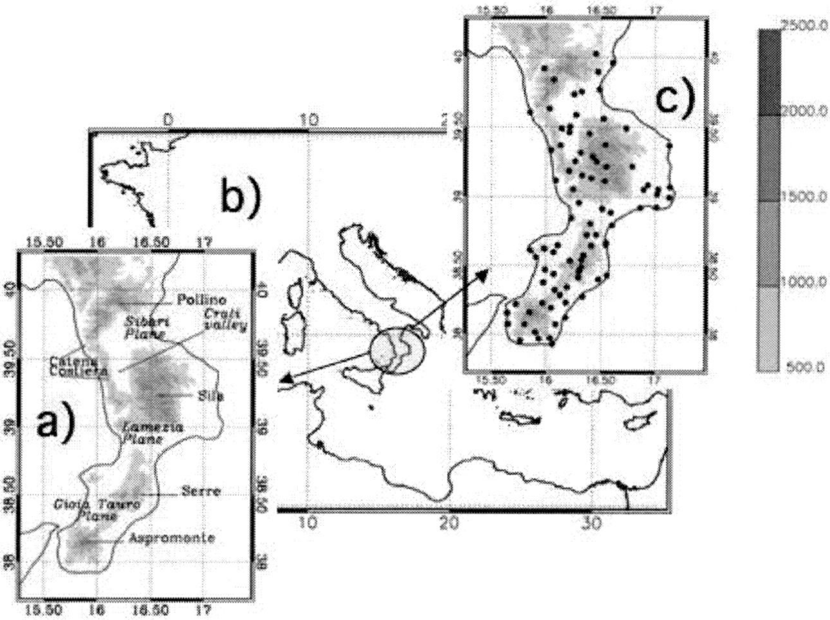

Figure 1. (a) Calabria features cited into the text; (b) Calabria region and its position in the central Mediterranean; (c) Black-filled circles are the surface stations of the regional meteorological network. Grey shading shows the orographic height (m).

The objective of this chapter is to show a methodology for studying the precipitation field over complex terrain by using the Calabria region as test-bed. First the climatology of the field is shown, and then the physical factors involved in heavy rainstorms are discussed. Finally, using the results of these studies, a more advanced, yet simple, forecasting technique is shown for improving the deterministic forecast.

2. PRECIPITATION CLIMATOLOGY

Figure 2 shows the average yearly precipitation for the period 1971-2000. These maps are derived from the data of the regional meteorological network (Figure 1) following the methodology reported in Federico et al. (2009). The orographic signature is quite clear: from north to south there are maxima corresponding to Pollino (1300 mm), Sila (1200 mm), Serre (1600 mm) and Aspromonte (1400 mm). The local maxima over Catena Costiera (1300 mm) could be underestimated because the mountain range is short in the west-east direction (less than 20 km) and raingauges are not very dense.

Generally, there is a rainfall gradient from west to east. Minimum precipitation near the coastline is less than 700 mm east of Pollino and Sila. The minima are a consequence of the shielding effect of the mountains because Mediterranean storms impinge Calabria from west to east and local orography enhances precipitation upwind of the mountains peaks reducing it downwind.

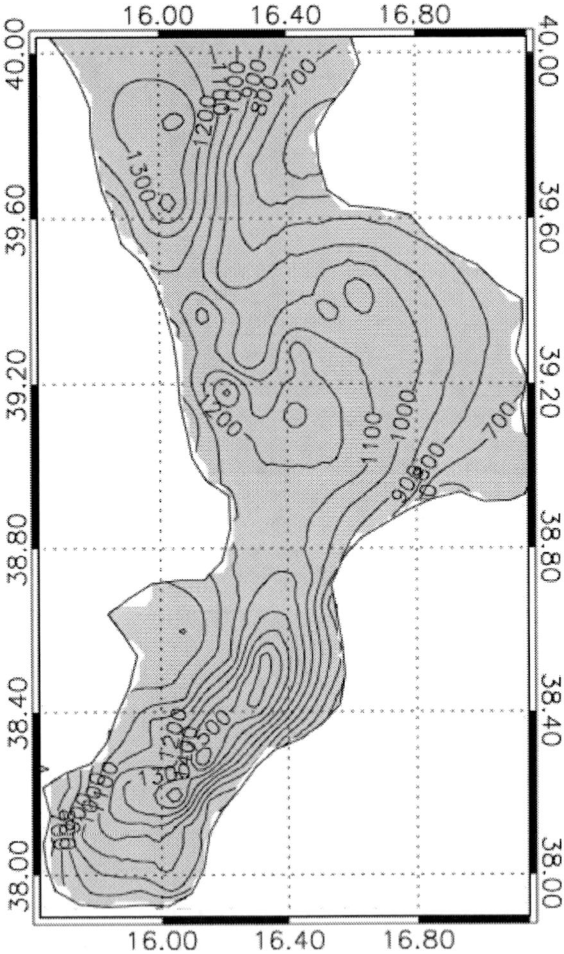

Figure 2. Average yearly precipitation for the period 1971-2000.

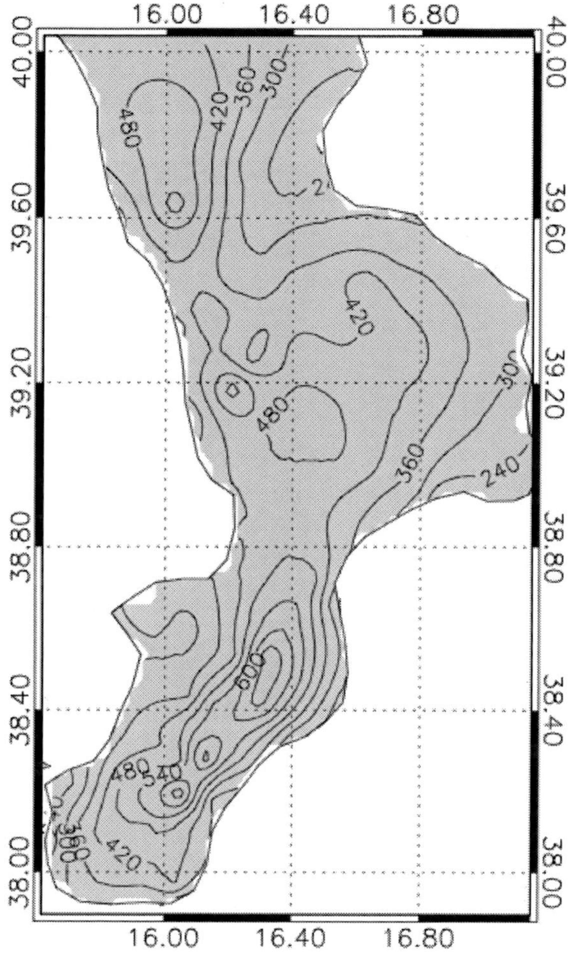

Figure 3(a). Average winter precipitation for the period 1971-2000.

By and large, the largest fraction of storms that affect Calabria are cyclones that develop on the lee of the Alps or that enter or originate into the Basin west of Calabria (Iberia, Rhone valley, Gibraltar) (Alpert et al., 1990; Trigo et al., 1999, 2002; Jansà et al., 2001; Romem et al., 2007; Federico et al., 2008a). The storms coming from north Africa and Balkans (the latter being storms that originate on the lee of eastern Alps and Balkans as "daughters" of European cyclones) are less frequent. The first class of cyclones usually travels from west to east or from northwest to southeast. They affect more the Tyrrhenian side of the peninsula in terms of rainfall recorded. The second class moves south of Calabria (African storms) or from northeast to southwest (storms from Balkans). During these storms, air masses impinge the Ionian coast of Calabria causing more precipitation on this side of the peninsula (Federico et al., 2008a).

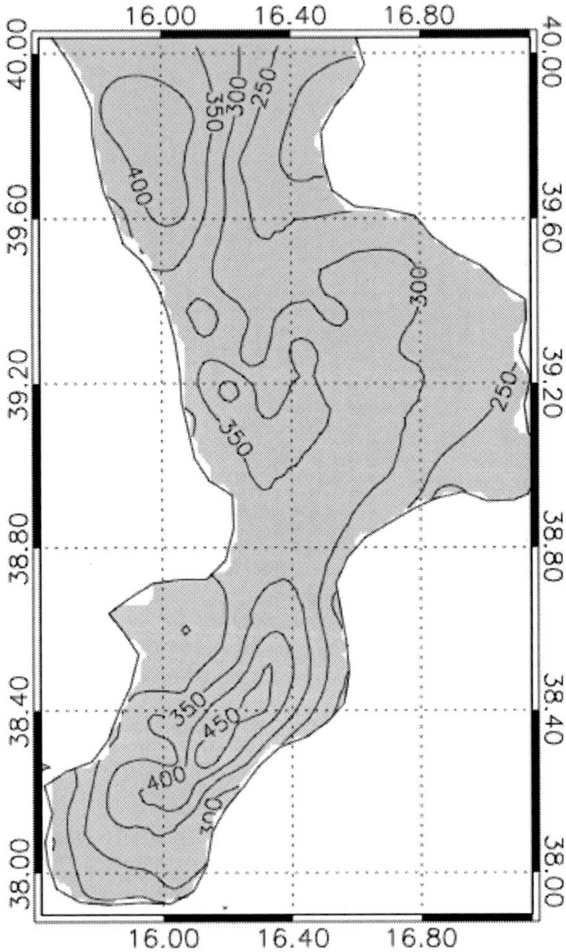

Figure 3(b). Average fall precipitation for the period 1971-2000.

An exception to this rule is on the east side of Serre and Aspromonte where precipitation is comparable to that recorded on the west side of the peninsula. These areas, in fact, are exposed to a larger number of cyclones compared to the areas east of Sila and Pollino, including those developing on the western Mediterranean. Noticeably, western Mediterranean storms force a cyclonic circulation that can extend over the Ionian Sea. Under this forcing, marine air masses are advected toward Calabria from south, south-east or east. Air masses coming from those directions are uplifted and produce more abundant precipitation over Serre and Aspromonte, compared to Sila and Pollino, because: (1) the orography is particularly steep, (2) air masses travel a longer fetch over the Ionian Sea, and (3) Sila and Pollino are partially sheltered by Serre.

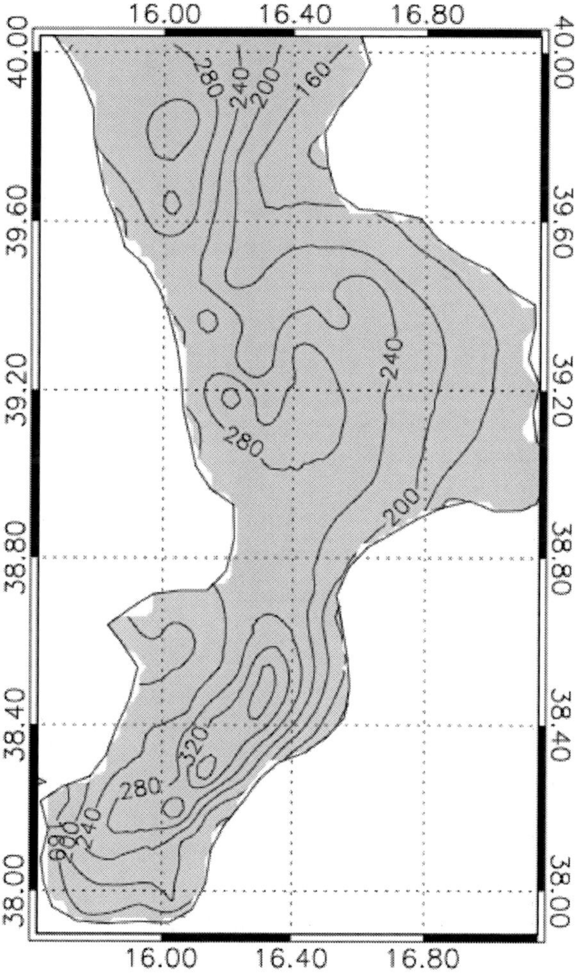

Figure 3(c) . Average spring precipitation for the period 1971-2000.

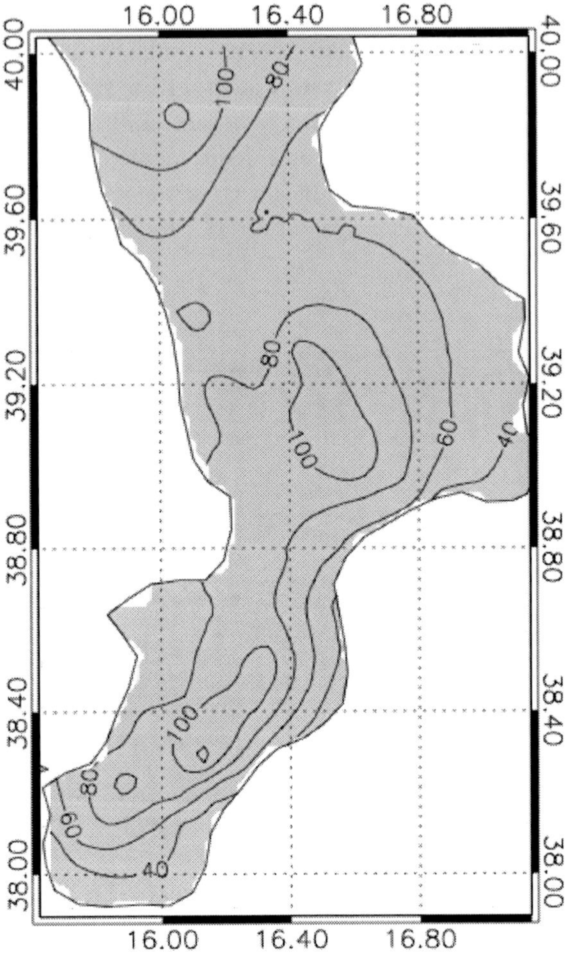

Figure 3(d). Average summer precipitation for the period 1971-2000.

Figures 3(a-d) show the averaged seasonal precipitation, i.e. (1) winter (December-February), (2) spring (March-May), (3) summer (June-August), and (4) fall (September- November). The maximum precipitation occurs in winter (Fig 3(a), maximum isoyeth 660 mm) followed by fall (Figure 3(b), maximum isoyeth 450 mm), spring (Figure 3(c), maximum isoyeth 360 mm) and summer (Figure 3(d), maximum isoyeth 120 mm).

The cold semester, from 1 October to 31 March, accounts for more than 80% of the total annual rainfall. This result clearly reflects the Mediterranean climate of Calabria, and in particular the effect of large scale pressure patterns.

All seasonal distributions reflect the pattern of the yearly average and the orographic effect is evident. These patterns are determined by the interaction between Mediterranean cyclones and the local orography.

It is interesting to focus on the incidence of heavy rainfall, shown in Figure 4, which is defined here as days with precipitation greater than or equal to 64 mm/day for the period 1971-2000. The total number of days with rainfall greater than 64 mm/day is less than forty (about one per year) in most locations. The eastern side of Calabria, however, is more exposed to heavy rainfall. From Figure 4, it is clear that the most vulnerable areas are Serre and Aspromonte. Furthermore, although the western coast of the peninsula has higher total annual precipitation as compared to the east coast, heavy rainy days are more frequent on the east side. This side of the peninsula is more exposed to a larger number of heavy rainy Mediterranean cyclones (Federico et al., 2008a).

An interesting feature of Figure 4 is the occurrence of heavy rainfall events in coastal stations of southeast Calabria and Marchesato (east of Sila). These events are associated to southerly, southeasterly and easterly wind and there are two main reasons for precipitation enhancement along the coastline: (1) the roughness increase from sea to land, and (2) the complex interaction between cyclones and orography.

The roughness increases from sea to land causing air masses convergence along the coastline, which forces upward motions of humid air coming from the sea. If suitable conditions are met, this forcing can trigger convection near the coast and intense precipitation occurs along the coastline (Romero et al., 1998).

Generally speaking, the complex interaction of orography and air masses produces two kinds of responses: (1) flow-around regime, and (2) flow-over regime. The orographically modified flow is determined by the environmental static stability, orographic height, and air velocity (Rotunno and Ferretti, 2001; Buzzi et al., 1998, Fantini and Davolio, 2001). Stratification effects on flow around topography are generally described in terms of the Froude number (Arya, 2001). The Froude number is given by $Fr=U/(NH)$ where U is the characteristic velocity of the flow, N is the Brunt-Väisälä frequency, and H is the topographical height. If $Fr \ll 1$, the stratification is strong and the flow-around regime occurs. If $Fr \gg 1$, the stratification is near neutral and the flow-over regime occurs. While for a particular storm one of the two regimes can be dominant, both regimes contribute to the rainfall patterns in a climatological sense. The adiabatic cooling of the flow over is the key factor for precipitation enhancement over the peaks and mountain slopes, while the flow around can produce convergence patterns along the coast. A typical case of this convergence occurs when a southeasterly main flow impinges onto the south-east coast, where it is deflected by Aspromonte and Serre. The upstream blocking produces convergence patterns along the coast, which forces upward motions and triggers convection.

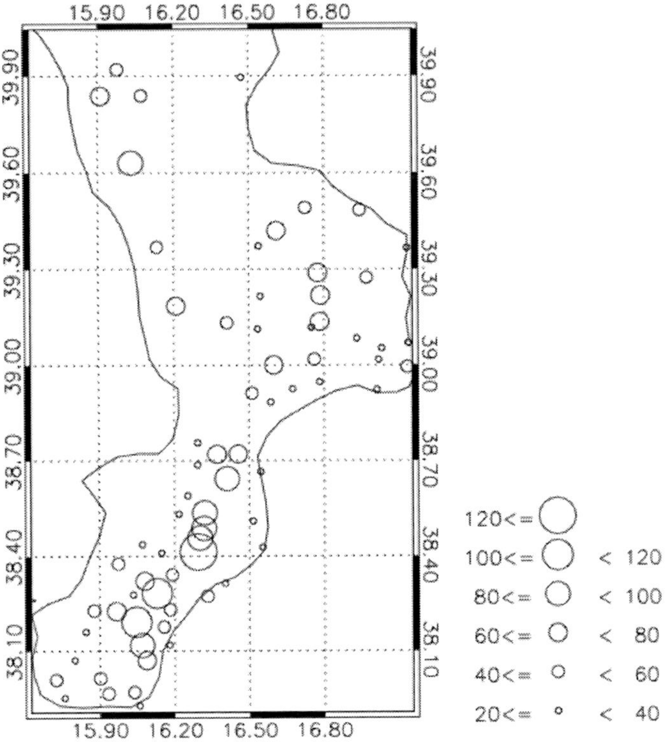

Figure 4. Number of days with precipitation greater than or equal to 64 mm/day for 1971-2000.

The seasonal analysis of heavy rainfall events reveals that the maximum number of cases occurs during winter followed by fall, spring, and summer. This is not surprising because: (a) winter has the maximum frequency of storm incidence over Calabria; (b) storms that occur in fall are more prone to heavy precipitation compared to spring because the sea is warmer in that season; (c) summer has the minimum incidence of cyclones. The spatial distribution of heavy rainfall does not show substantial differences among seasons, except during summer when the western side of the peninsula receives more heavy rain.

3. Physical Factors Involved and Sensitivity to the Large Scale Forcing

From a climatic viewpoint, as discussed in Section 2, the orography, the presence of the Mediterranean Sea, and the large scale atmospheric forcing play a crucial role in rainfall events over Calabria. On 10-12 December 2003, a particularly interesting event occurred when an unusually high rainfall fell over the country. In Sections 3 and 4, using the event as a case study, the roles of the physical factors, i.e. the orography and surface latent heat flux, and of the large scale forcing are investigated. Even though the investigation is carried out over a particular case study, the factors considered are of general importance for heavy precipitation events in the Mediterranean Basin (Fehlmann and Davies, 1999; Fehlmann et al. 2000; Homar et al. 2001, 2002; Martin et al. 2007; Moscatello et al, 2008; Krichack et al. 2007; Siccardi, 1996; among others).

3.1. The Role of Orography and Surface Latent Heat Flux

Figure 5 shows the raingauges accumulated precipitation from 00:00 UTC on 10 December 2003 to 00:00 UTC on 13 December 2003, which were objectively analyzed by a Cressman-type scheme with 0.15° search radius, that smears the 660 mm maximum measured value. During the storm, the wind direction was mainly from the southeast and there was a partial shielding caused by the mountain ranges that led to more precipitation on the Ionian coastal areas. Moreover, accumulated rain was greater over the main peaks (Sila, Serre, and Aspromonte) due to the orographic uplift.

Figure 6(a) shows the low level ECMWF (European Centre for Medium-range Weather Forecast) operational analysis at 12:00 UTC on 10 December 2003. A low pressure pattern was centred over Algeria and warm humid air masses were advected onto the Calabrian east coast. During the following 12 hours, intense rainfalls were reported in several villages of the Ionian coast.

For the same time as that in Figure 6(a), Figure 6(b) shows the upper level analysis. For the extra tropics, PV > 3 PV units (PVU[2]) are inferred to represent stratospheric air masses because of the high values of static stability in this portion of the atmosphere. PV = 1.5-3 PVU represent air that originated near the tropopause. PV < 1.5 PVU represent tropospheric air masses (Krichack et al., 2004, 2007). Positive PV anomalies are often found in areas with downward propagation of stratospheric air masses (Santurette and Georgiev, 2005), and following the classification described above, the height of the 1.5 PVU surface is usually assumed as the height of the dynamical tropopause.

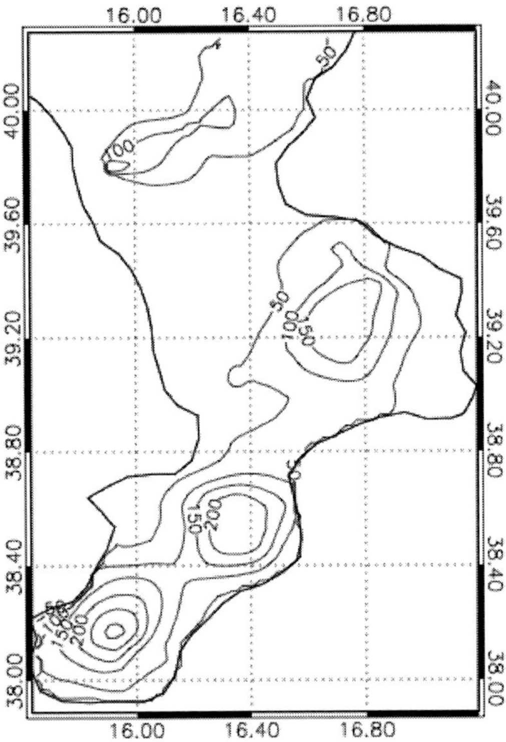

Figure 5. Total precipitation (mm) from 00:00 UTC on 10 December 2003 to 00:00 UTC on 13 December 2003. Contours from 50 to 300 mm every 50 mm (from Federico et al., 2008b).

[2] 1 PVU=10^{-6} m^2 s^{-1} K kg^{-1}.

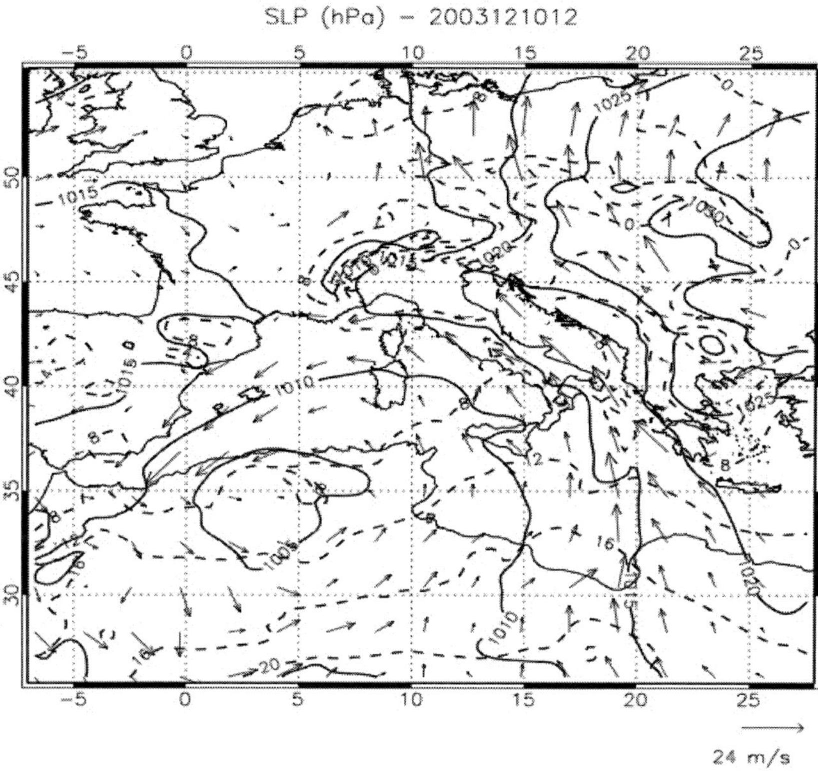

Figure 6(a). Sea level pressure (hPa, solid contours), 925 hPa temperature (°C, dashed contours) and 925 hPa winds. The map is valid at 12:00 UTC on 10 December 2003 (adapted from Federico et al., 2008b).

During the following day, the cyclone moved eastward and the pressure minimum reached the Libyan coast at about 12:00 UTC on 11 December. Then it moved northeastward crossing Calabria and Sicily. This evolution is well illustrated by Figures 7(a and b).

Figure 7(b) shows also the upper level cut-off associated with the surface low and the elongated tropospheric intrusion of stratospheric air masses over Africa.

It is worth noting that winds over the Ionian Sea intensified throughout the cyclone evolution and advected moist air masses toward the Ionian coast. In addition, the height of the 1.5 PVU surface in Figure 7(b) is lower than that in Figure 6(b), and its minimum height is closer to the central Mediterranean basin. This meteorological situation determines a more direct upper level forcing of the circulation over the target area.

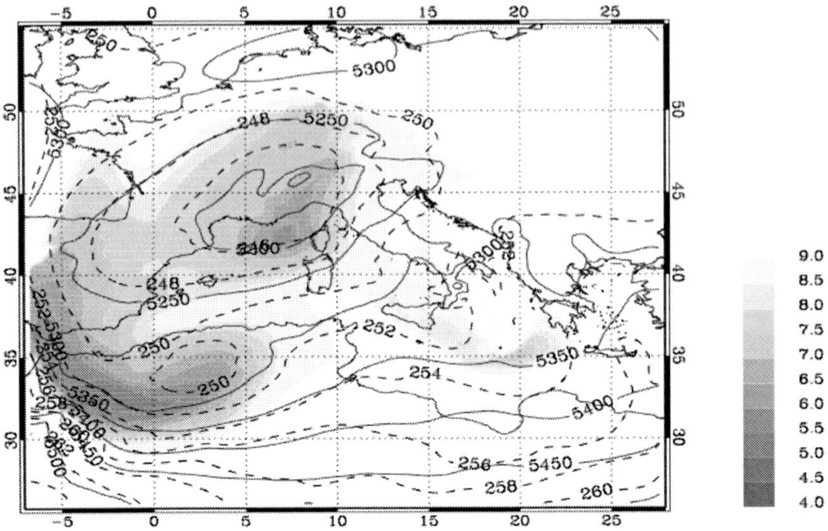

Figure 6(b). 500 hPa geopotential height (m, solid contours), 500 hPa temperature (K, dashed contours) and 1.5 PVU surface height (km, filled contours). The map is valid at 12:00 UTC on 10 December 2003 (adapted from Federico et al., 2008b).

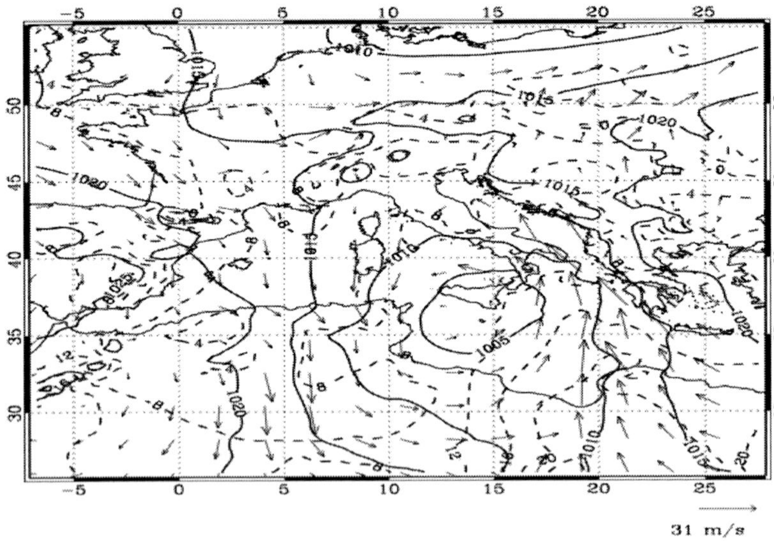

Figure 7(a). Sea level pressure (hPa, solid contours), 925 hPa temperature (°C, dashed contours) and 925 hPa winds. The map is valid at 00:00 UTC on 12 December 2003 (adapted from Federico et al., 2008b).

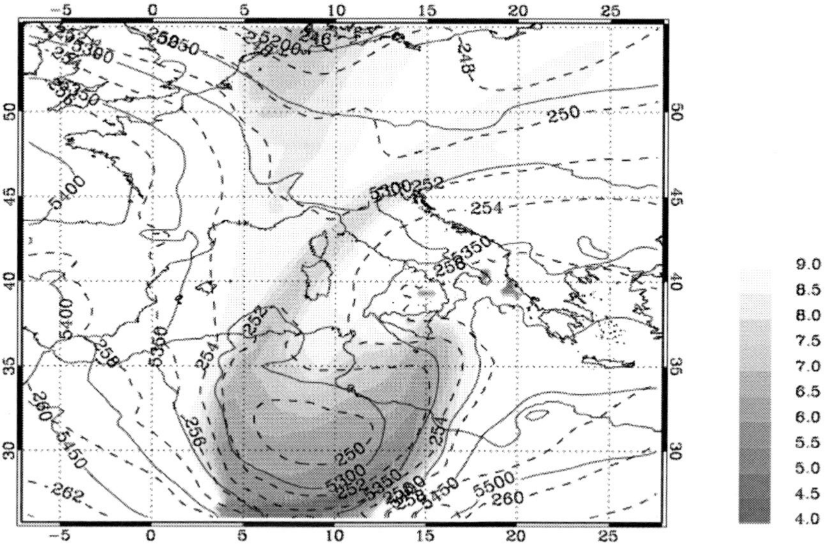

Figure 7(b). 500 hPa geopotential height (m, solid contours), 500 hPa temperature (K, dashed contours) and 1.5 PVU surface height (km, filled contours). The map is valid at 00:00 UTC on 10 December 2003 (adapted from Federico et al., 2008b).

Calabrian orography and surface latent heat flux were important physical factors for this event because the surface wind pattern was mainly from the southeast and advected relatively warm humid air masses toward the peninsula. To study this issue, the factor separation technique, proposed by Stein and Alpert (1993), was applied in order to isolate the role of each factor and their combined effect. When more than one factor is considered, it is not sufficient to compare a simulation done by removing some physical effect from the control simulation (run f_0) because the contribution associated with the interaction of two (or more) factors must be considered. A complete factor separation for n factors requires 2^n simulations, and thus the results of four simulations are shown thereafter. Table 1 contains the four simulations and the algebraic combinations used to isolate the effects of the two physical factors on the precipitation field.

In this chapter, all model simulations were performed using the RAMS model (Pielke, 2002). The physical setting of the RAMS model is summarized in Table 2. The setting is identical to the operational implementation currently available for southern Italy.

The initial and dynamic boundary conditions were derived from the ECMWF 12:00 UTC operational deterministic analysis/forecast cycle issued on 9 December 2003[3].

Table 1. Simulations performed to show the effects of the physical factors involved in the December 2003 case study

Simulation run	Physics	Factor separation
f_0	Full physics	$f_0+f_1-(f_2+f_3)$= combined effect
f_1	No Calabria orography, no surface latent heat flux	/
f_2	Calabria orography, no surface latent heat flux.	f_2-f_1= role of the calabrian orography
f_3	No Calabria orography, surface latent heat flux.	f_3-f_1= role of the surface latent heat flux

The control run is able to represent the main features of the precipitation field in terms of the total amount and horizontal gradients in the north-south and west-east directions (Figure 8(a)). Considerable precipitation was predicted over the Sila and Pollino mountain ranges even though they were underestimated. It is important to note that modelled rainfall often overestimates precipitation upwind of the main peaks (Romero et al. 1998; Pastor et al. 2001; Mass et al. 2002; Federico et al. 2003a) for severe weather and for high horizontal resolutions as those adopted in this chapter. This is clearly evident over Aspromonte where RAMS overestimated rainfall on the upwind side of the mountain peak (i.e. east and south of the main peak) and underestimated precipitation leeward.

Figure 8(b) shows the total precipitation for the run in which Calabria orography was flattened to 1 m height (forecast 3, f_3) but with surface latent heat flux activated. Approximately, it accounts for the role of surface latent heat fluxes (f_3-f_1 in Table 1) because precipitation for experiment f_1 was much less than that simulated in f_3. Rainfall along the Ionian coast was confined in a short distance near the coastline because the surface roughness changed along

[3] Analyses are often used as initial and boundary conditions to study past weather events. In this and in the following sections the operational analysis/forecast cycle of the ECMWF is used in place of the analysis to stress the usefulness of the results in an operational context.

this line and forces wind convergence and upward motions. The environment was wet enough that air masses were able to reach their Level of Free Convection (LFC), which was as low as 950 hPa before the initiation of convection. The role of surface latent heat flux was not negligible, nevertheless total precipitation is largely reduced compared to the control case suggesting the important contribution of other factors (as orography flux or the synergetic effect of orography and surface latent heat flux) to the total precipitation.

Table 2. RAMS model physical settings

Physical option	Description
Parametrized cumulus convection	Modified Kuo scheme to account for updraft and downdraft (Molinari and Corsetti, 1985).
Explicit precipitation parametrization	Bulk microphysical model which prognoses cloud water, rain, ice crystals, aggregates, graupel and hail (Walko et al., 1995).
Subgrid mixing	The turbulent mixing in the horizontal directions is parameterized following Smagorinsky (1963), which relates the mixing coefficients to the fluid strain rate and includes corrections for the influence of the Brunt-Vaisala frequency and the Richardson number (Pielke, 2002). Vertical diffusion is parameterized according to the Mellor and Yamada (1982) scheme, which employs a prognostic turbulent kinetic energy.
Exchange between the surface, the biosphere and the atmosphere.	LEAF-3 sub-model (Walko et al., 2000). LEAF includes prognostic equations for soil temperature and moisture for multiple layers, vegetation temperature and surface water including dew and intercepted rainfall, snow cover mass and thermal energy for multiple layers, and temperature and water vapour mixing ratio of canopy air.
Radiation scheme	A full-column, two-stream single-band radiation scheme is used to calculate short-wave and long-wave radiation (Chen and Cotton, 1983). The Chen and Cotton scheme accounts for condensate in the atmosphere, but not for specific optical properties of ice hydrometeors.

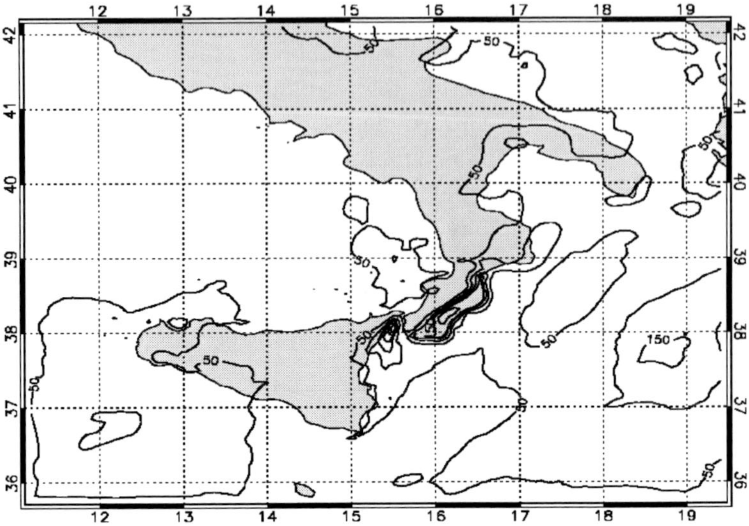

Figure 8(a). Total precipitation accumulated for the RAMS control simulation for the same time interval as Figure 5. Contours from 50 mm to 250 mm, every 50 mm. Contour labels for 50 mm and 150 mm (from Federico et al., 2008b).

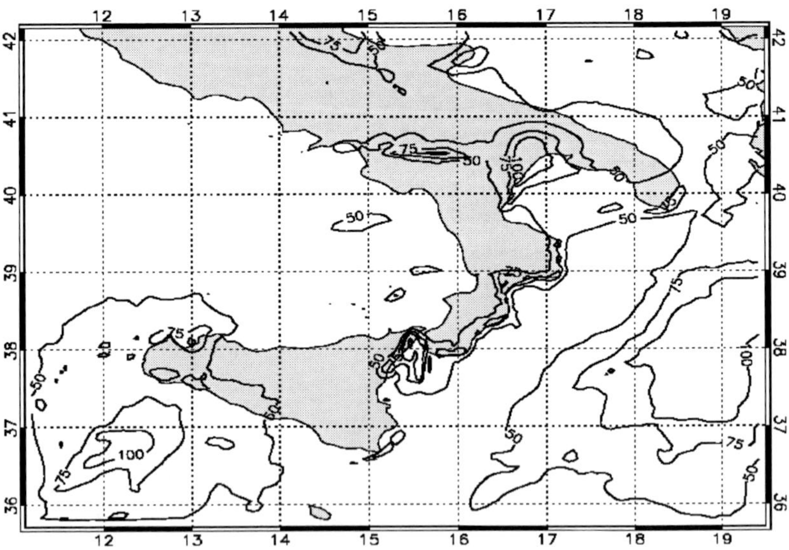

Figure 8(b). Total precipitation accumulated for the simulation with no calabrian orography (f_3) for the same time interval as Figure 5. Contours from 50 mm to 100 mm every 25 mm (from Federico et al., 2008b).

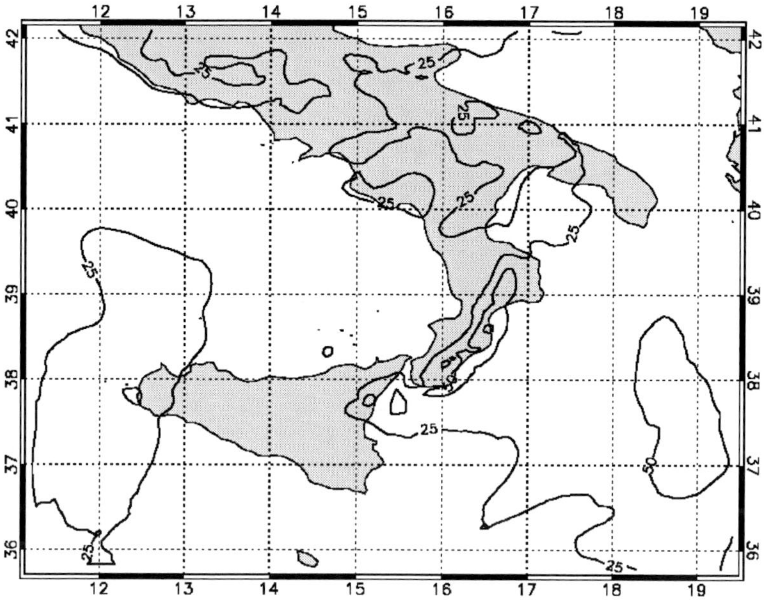

Figure 8(c). Total precipitation accumulated for the simulation with no surface latent heat flux (f_2) for the same time interval as Figure 5. Contours for 25 mm and 50 mm (from Federico et al., 2008b).

Figure 8(c) shows the total precipitation for the run with no surface latent heat fluxes (f_2) but with the actual Calabrian orography. Approximately, it accounts for the role of orography (f_2-f_1 in Table 1) because in simulation f_1, precipitation for was much less than that in simulation f_2. In this case, the effect of topography was apparent because precipitation maxima were windward along the main peaks. However, total precipitation was largely reduced as compared to the control case, showing the contribution of other factors (as surface latent heat flux or the synergetic effect of orography and surface latent heat flux).

Figure 8(d) shows the synergetic effect of the orography and surface latent heat flux (f_0-f_3-f_2+f_1 in Table 1). Humid air masses were advected from the Ionian Sea toward Calabrian east coast where they interacted with the steep mountain ranges determining rainfall. This situation persisted for two days and contributed to the unusually large amounts of precipitation recorded for this event. The orography and surface latent heat flux synergetic effect gives by far the most important contribution to the total precipitation over southern Calabria compared to the previous effects analyzed, and accounts for most of the precipitation simulated in the control run (f_0).

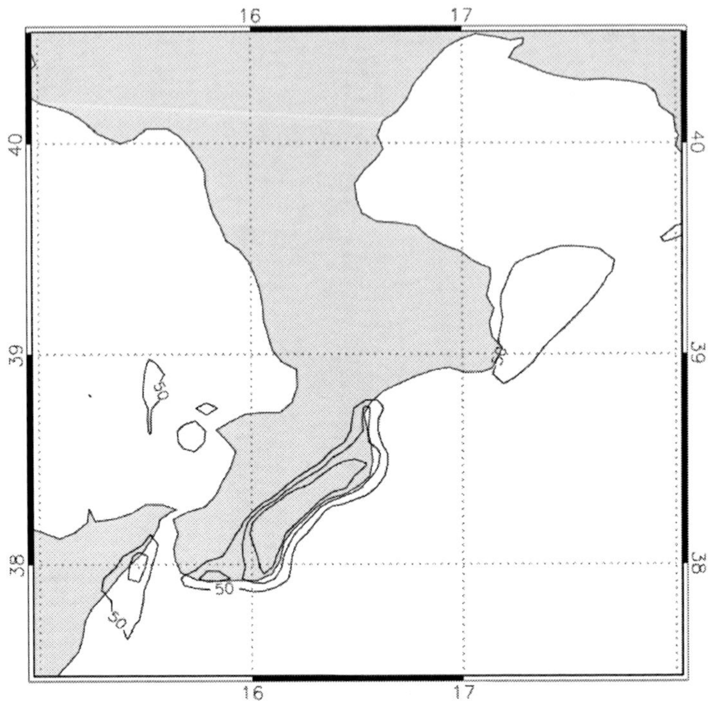

Figure 8(d). Total precipitation accumulated for the combined effect (f_0-f_3-f_2+f_1) for the same time interval as Figure 5. Contours for 50 mm, 100mm, and 150 mm. The figure has been enlarged over the target area to better see the contours (adapted from Federico et al., 2008b).

In conclusion, the boundary factor (orography) and the physical factor (surface latent heat flux) were fundamental mesoscale ingredients because they determined intense and abundant rainfall over Calabria, mainly by their synergetic effect. The following section explores the storm sensitivity to the large-scale forcing.

3.2. Sensitivity to the Synoptic Scale Forcing

The strategy adopted to study the storm sensitivity to the synoptic scale forcing[4] consists of generating different "scenarios", compatible with the analysis/forecast error, with respect to a reference run (hereafter also referred

[4] The technique shown in this paragraph is also useful to evaluate the storm predictability trough the ensemble spread (Molteni et al., 2001).

to as control). In this approach it is assumed that the model is perfect and that the major error sources are associated with deficiencies in the knowledge of initial conditions ingested into the mesoscale model, a real problem in the Mediterranean.

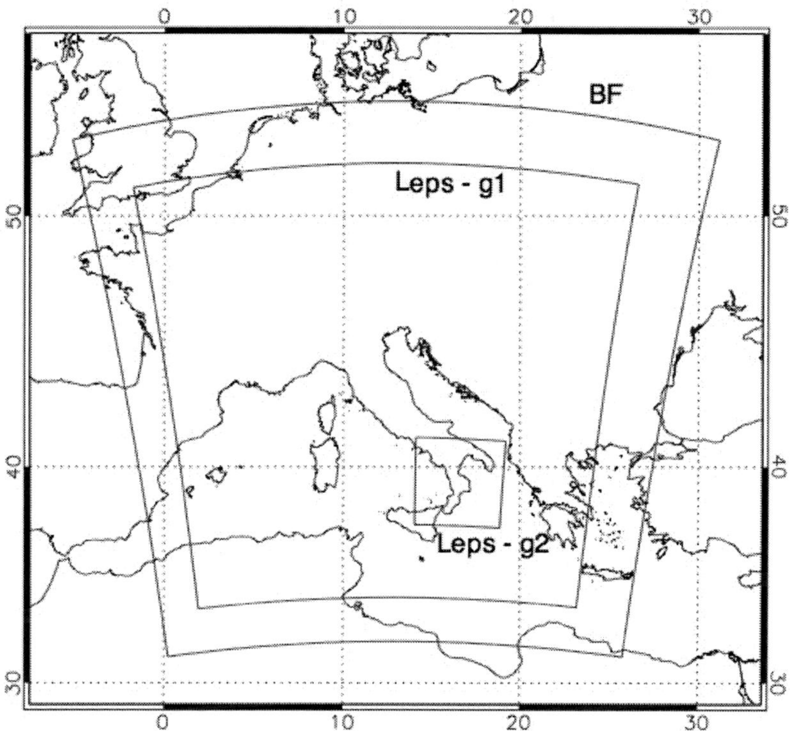

Figure 9. Domains of RAMS-BF (horizontal resolution: 50 km) and RAMS-LEPS (horizontal resolutions 30 km and 6 km for g1 and g2, respectively).

In the extra-tropical atmosphere intense cyclones often develop under upper-tropospheric jet streams that are often characterized by anomalously high values of PV. Potential vorticity is conserved along the flow on a isentropic surface under adiabatic and frictionless motion (Hoskins et al. 1985). Because of its conservative property, PV was used to trace areas with specific air mass characteristics and was applied in the analysis of atmospheric processes. The positive PV anomalies influence the timing, amplitude, and location of intense surface cyclones and heavy precipitation events by upper level forcing (Fehlman and Davies, 1999), and accurate specification of the tropospheric PV distribution is necessary for the successful prediction of hazardous weather events (Romero et al. 2005). In other words, the PV can be used as the key variable for accounting the multitude of parameters characterizing a synoptic-scale storm.

A LEPS approach was used to generate PV perturbations because it provides different scenarios compatible with current analysis/forecast errors. RAMS-LEPS methodology is detailed in the following two points:

5. Five sets of perturbed initial and boundary conditions were generated by a hierarchical clustering technique applied to the results of a Brute Force (BF) run performed by the RAMS model with 50 km horizontal resolution. The RAMS-BF was initialized by the ECMWF-EPS[5] analysis and forecast cycle issued at 12:00 UTC on 9 December 2003, and was used to compute derived variables as PV or the height of the dynamical tropopause. The 51 runs of the RAMS-BF, driven by the 51 global predictions of the ECMWF-EPS, were gathered by the following methodology:
    - We started with 51 clusters where each RAMS-BF member belongs to one cluster.
    - From t_0+24 h to t_0+84 h, where t_0 is the initial simulation time, the space-averaged quadratic distances between all RAMS-BF members for the height of the dynamical tropopause was computed. The space-averaged quadratic distance between RAMS-BF members k and l, at time t, is given by:

$$d(t) = \frac{1}{N}\sum_{i=1}^{N}(hgt_k(i,t) - hgt_l(i,t))^2 , \qquad (1)$$

      where hgt_k and hgt_l are the heights of the dynamical tropopause for the members k and l, respectively, and the summation spans over all grid points inside the central Mediterranean basin (5E-20E, 35N-45N). The distance between members k and l is determined by averaging $d(t)$ over all available model outputs from t_0+24 h to t_0+84 h. It is stressed that the time window and the target area of the clustering technique focus the result of the methodology to the second and third forecast-day and to the central Mediterranean.

[5] When the storm occurred, fifty perturbed members and the unperturbed forecast composed the ECMWF-EPS. Its horizontal resolution was about 90 km over the target area.

- The two clusters with minimum distance were grouped together.
- Next, distances between different clusters were computed by averaging distances between all cluster members.
- The previous two steps were repeated until only five clusters remains. The scenarios number (five) was selected by trial and error considering two basic requirements: LEPS should have a proper representation of different scenarios simulated by RAMS-BF, and; the computational time should be short enough for the practical implementation of the methodology.
- One representative member (RM) was selected for each of the five clusters by minimizing the ratio between the distance from its own cluster members and the distance from members belonging to other clusters.

6. For each of the five RMs, a high resolution RAMS forecast, with 6 km maximum horizontal resolution, was performed to resolve local scales which have a high impact on local weather and are not represented properly in the ECMWF-EPS and in the RAMS-BF because of their coarse horizontal resolutions.

These five simulations ("scenarios") form the RAMS-LEPS. Figure 9 shows the horizontal domains used by RAMS-BF and RAMS-LEPS. The former has one grid with 50 km horizontal resolution, while the latter has two nested grids with 30 km and 6 km horizontal resolutions (two-way nesting). Thirty vertical levels are used for both configurations spanning the region 0-17 km.

Table 3 contains a summary of the application of the clustering technique for this case study. From here onwards, the RAMS-BF or RAMS-LEPS member is indicated by the number of the corresponding ECMWF-EPS member, whose fields drive the RAMS simulation. Cluster 1 (RM 06) and Cluster 2 (RM 09) account for 90% of the total RAMS-BF members. Cluster 3 (RM 07) has three members and represents 6% of the total ensemble population. Cluster 4 (RM 10) and Cluster 5 (RM 44), each with one member, represent 4% of the total ensemble population.

To study the sensitivity of the storm to the large-scale forcing, it is useful to consider the height of the dynamical tropopause simulated by the RMs of the five clusters and by the unperturbed forecast on 12 December at 00:00 UTC (Figure 10). All RMs show the PV anomaly for this event, even if there are differences between RMs. Hereafter RM 06 was chosen as the reference because it is the RM of the most populated cluster and the unperturbed forecast belongs to its cluster.

Table 3. Result of the application of the hierarchical clustering technique for the December 2003 case study

Cluster #	Cluster population	Representattive member (RM)
1	26	6
2	20	9
3	3	7
4	1	10
5	1	44

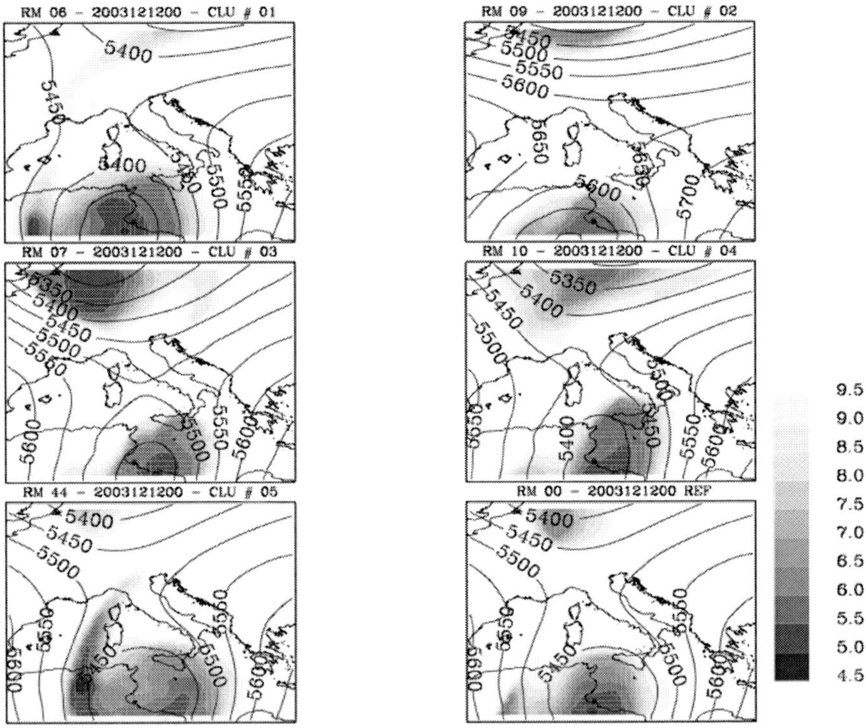

Figure 10. Height of the dynamic tropopause (km, filled contours) and 500hPa geopotential height (m, black contours) simulated at 00:00 UTC on 12 December 2003 by the representative members (RM) of the five clusters (the RM number is in the upper left corner of each image). The lower right image (RM 00) is the unperturbed forecast (from Federico et al., 2008b).

The height of the dynamical tropopause for RM 09 (cluster 2) over Africa is higher as compared to RM 06 and its pattern is slightly shifted to the east. Model output shows that the lower height of the dynamical tropopause corresponds to less intense surface winds, lower rain rate, and less accumulated rain at the end of the forecast (Figure 11).

The RM 07 upper level cutoff and the minimum height of the dynamical tropopause over Africa are shifted to the east. This was caused by a faster movement of the RM 07 cyclone from west to east with important consequences on the precipitation field (Figure 11). As expected, RM 07 has less rainfall because the cyclone and the surface wind field were less stationary over Calabria as compared to RM 06. The eastward transition speed of the cyclone is a key feature of this storm. Faster (slower) transition corresponds to shorter (longer) interaction of surface humid air masses with the steep Calabrian orography and less (greater) rainfall.

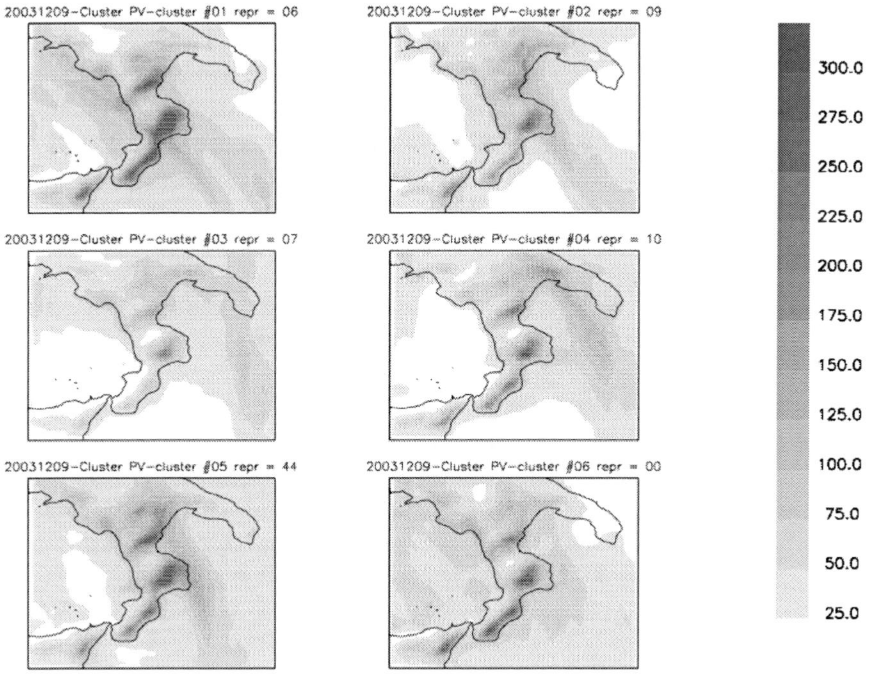

Figure 11. Total rainfall (mm) simulated by the representative members (RM). The RM number is in the upper right corner of each image. The lower right image (RM 00) is the unperturbed forecast (from Federico et al., 2008b).

The RM 10 height of the dynamical tropopause pattern is similar to that of RM 07 and is shifted to the east as compared to RM 06, however its precipitation is larger (Figure 11) than RM 07. As shown in Figure 10, the shapes of the upper level cutoff and of the dynamical tropopause are different and extend more northward for RM 10 as compared to RM 07. Model output analysis reveals that this determines precipitation that lasts longer over Calabria. Similar considerations apply to RM 44.

The unperturbed forecast shows a similar pattern of the dynamical tropopause and upper level cutoff to RM 06 but it is again shifted to the east. This has consequences on the accumulated rainfall for the two simulations (Figure 11). Again, RM 06 has more abundant precipitation because the simulated cyclone is more stationary over Calabria and this produces larger rainfall.

The RAMS-LEPS performs well for this case and, with the exception of RM 07, the RMs are able to describe the development of the cyclone and the associated rainfall pattern. The use of five scenarios gives a satisfactory representation of the storm evolution. Figures 10 and 11 show the storm sensitivity to: (1) different values of the height of the dynamical tropopause (RM 06-09), (2) cyclone progression from west to east (RM 06-07), and (3) different shapes of the height of the dynamical tropopause (RM 06, RM 07, RM 10, RM 44).

4. A Forecasting Technique Using Water Vapour Images

An interesting question arising from the study of the cyclone sensitivity to the synoptic scale forcing (Section 3) is that: which is the best scenario? This question is the basis for a forecasting technique based on the subjective comparison between METEOSAT[6,7] water vapour images and RAMS-LEPS pseudo water vapour images. These maps can be compared for choosing the representative member that is following better the real atmospheric behaviour, or for discarding scenarios not representing the real situation.

---

[6] In this chapter only METEOSAT first-generation water vapour images are used.

[7] In this chapter the comparison between the model pseudo water-vapor images and the METEOSAT images is subjective. The same comparison can be made quantitative by using the inversion property of the potential vorticity as reported in several studies (see Santurette and Georgiev, 2005 for a general review).

The METEOSAT water vapour images are derived from radiance measurements in the spectral region centred around 6.3 μm and provide a measure of the integrated water vapour content above 500 hPa with a horizontal resolution of about 8 km x 5 km (Fisher et al., 1981). Dark band on these images in midlatitudes are an indicator of significant in situ descent (Menzel et al., 1998) and concomitant high PV values. Indeed there is a strong resemblance between these images and the PV distribution on troposphere transcending isentropic surfaces (Appenzeller and Davies, 1992; Appenzeller et al., 1996).

In cloud-free region there is a reasonable quantitative relationship between the temperatures on the water vapour isosteric surface of 75 mg/kg and the corresponding water vapour image, and model pseudo water-vapour surfaces are easily computed as the temperature of the isosteric surface (Ramond et al., 1981, Appenzeller et al., 1996). In cloud-free regions, the comparison between the model pseudo water-vapour image and the satellite image can be used to subjectively evaluate the quality of the forecast.

Figure 12(a) shows the METEOSAT 7 water-vapour image at 00:00 UTC on 12 December 2003. The upside-down comma-shaped dark band extending from North Africa to North Corsica clearly shows the presence of the PV filament associated with the storm. The comparison between the pseudo WV image for RM 06 (Figure 12b) and the corresponding METEOSAT scene shows a good agreement. In particular, the dark band associated with the RM 06 PV filament has the same shape and is correctly co-located with the real PV structure, showing that RM 06 is giving a very good forecast. Some minor differences arise, as the location of the dark band over the Ligurian Sea, but RM 06 represents the structure of the PV filament and its location.

The comparison between WV and pseudo WV images for RM 07 (Figure 12c) shows that this scenario is unsatisfactory for both the location and shape of the dark band. The PV filament is to the east of the real METEOSAT image, showing that the storm movement toward the east is faster than the real movement. For example the western part of the RM 07 PV filament crosses Sardinia, while it is east of the Balearic Islands in the real image. Moreover the shape of the PV structure is different from the real METEOSAT image. For example the distance between the westernmost and easternmost parts of the PV filament is lower for RM 07 than for the real image.

From the above subjective analysis it is evident that the RM 06 is giving a better representation of the real atmospheric evolution than RM 07 and the precipitation field forecast by RM 06 (Figure 11) is more reliable than that of RM 07.

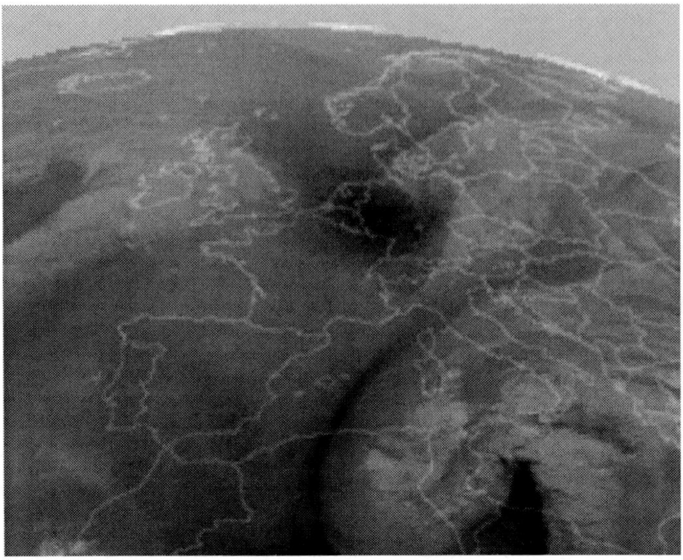

Figure 12(a). METEOSAT 7 WV image at 00:00 UTC on 12 December 2003. Copyright © 2011 EUMETSAT.

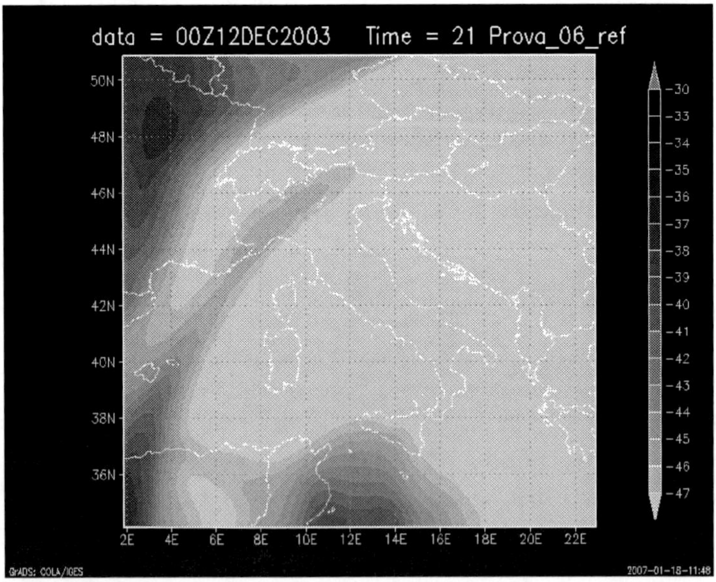

Figure 12(b). RAMS pseudo WV image at 00:00 UTC on 12 December 2003 for the RM 06 forecast. The image has been derived from the RAMS-LEPS forecast. The plotted field is the temperature (°C) of the 75 mg/kg isosteric surface.

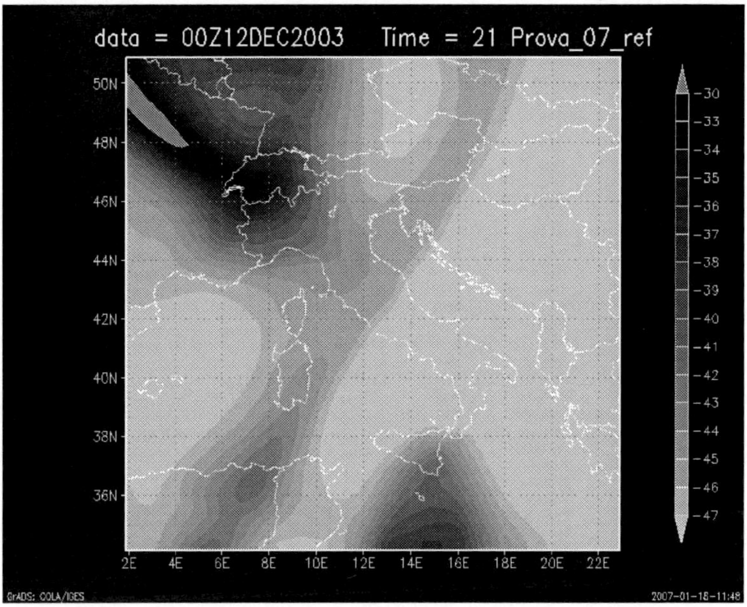

Figure 12(c). RAMS pseudo WV image at 00:00 UTC on 12 December 2003 for the RM 07 forecast. The image has been derived from the RAMS-LEPS forecast. The plotted field is the temperature (°C) of the 75 mg/kg isosteric surface.

The agreement between the METEOSAT WV image and others RMs water vapour images is between that of RM 06 and RM 07. From the above analysis, it is concluded that the member predicting a comparatively lower rainfall over the target area is the RM showing the largest differences from the real behaviour of the atmosphere. So, the application of this methodology can be useful to refine the deterministic forecast of the previous section.

This simple technique can be more useful when the RMs show a larger disagreement between scenarios, as reported in Federico et al. (2007) for a moderate storm occurred in May 2002.

CONCLUSION

This chapter discusses several aspects of the precipitation field over Calabria peninsula, southern Italy. Calabria offers an interesting test-bed for mesoclimatic studies, because of its orography and geographical position. The country is located in the centre of the Mediterranean Basin, which is among the most cyclogenetic areas of the world, and is characterized by elevated mountains near the coast. The physiographical characteristics of the country are typical of several areas in the Mediterranean and in the world thereby giving to the results presented a more general application.

Both from a climatic viewpoint and for single storms, the physiographical characteristics of the country play a crucial role for the precipitation field.

From a climatic viewpoint, rainfall spatial and temporal variability over Calabria is the result of the following three main factors:

(1) The orography enhances the precipitation on the upwind side (the west side of Calabria) and over the main peaks, while reduces it downwind (the east side of Calabria).
(2) The synoptic scale conditions give to the precipitation field a clear seasonal dependence. The rainiest season is winter followed by fall, spring and summer.
(3) The sea surface temperature is higher in fall than in spring increasing the surface latent heat flux during this season. This feature determines more abundant precipitation in fall as compared to spring.

Although orography is important for the precipitation enhancement over slopes and mountainous peaks, heavy rainfall is not uncommon near the coast. Two main reasons for precipitation enhancement have been highlighted: (1) the increase in roughness from sea to land, and (2) the interaction between the cyclones and orography determines convergence patterns along the coastline.

In the second part of this chapter, the RAMS model was used to study an intense and high impact storm that occurred over Calabria on 10-12 December 2003. The storm was characterized by the entrance of an Atlantic upper-level trough associated with a deep surface low that advected humid marine air toward the steep Calabrian orography. The mountain ridges favoured landslides, which were reported in several villages and damage to property was extensive. This storm was chosen among others because it is a prototype of intense rainfall events over Calabria and can be used to show the key factors involved in such events, both at the synoptic and mesoscale.

The results of this case study show that mesoscale factors play a major role during heavy rainfall events over the country. Indeed, results from the factor separation technique show that Calabrian orography and surface latent heat flux from the Mediterranean Sea contributed sizeably to the total precipitation amount, because their synergetic action accounts for the largest part of the total precipitation. This is expected by the examination of the synoptic scale forcing that advected humid and relatively warm, i.e. from the south, air masses toward the Calabrian east coast.

The synoptic scale environment was crucial for this rainstorm. To study this point, different scenarios compatible with the analysis and forecast errors were generated by a LEPS, and were used to show the rainfall variability to changes in the synoptic scale environment. The LEPS hierarchical clustering uses the height of the dynamical tropopause as the discriminating variable and, in this context, the PV is used as the variable "marking" the cyclone characteristics. The variability of the precipitation field for different scenarios shows the sensitivity of the precipitation field to the synoptic scale conditions.

The approach used for the December 2003 case study defines a general technique to improve the deterministic forecast. The technique relies on the subjective comparison between the METEOSAT WV images and model pseudo water vapour images. Different scenarios compatible with the analysis and forecast errors are generated by the LEPS. A model pseudo water-vapour image is computed for each representative member of the scenarios. This image is simply the temperature of the 75 mg/kg isosteric surface. The subjective comparison between the METEOSAT WV image and the model pseudo WV images is used to check the reliability of different scenarios. It is important to stress that LEPS resolve the local scale, which are very important for the precipitation field. In the current operational implementation at CRATI/ISAC-CNR, the LEPS has 6 km horizontal resolution over Calabria. It is also noticed that, by choosing the LEPS spatial-temporal domain of the clustering technique, the methodology can be targeted for different time windows and spatial domains.

For the case study reported in this chapter, the comparison between the METEOSAT WV image and the LEPS pseudo WV images gives less reliability to the forecast of the only scenario predicting comparatively lower precipitation. By this approach the determinist forecast, which predicted unusually heavy rainfall over Calabria, would have been considered highly reliable.

ACKNOWLEDGMENTS

This work was realized in the framework of the project MAPVIC - "Modellistica Agrometeorologica di Precisione per la VIticoltura Calabrese" funded by the Calabria Regional Government. I acknowledge my colleague

Elenio Avolio who shared with me most of the research discussed in this chapter.

REFERENCES

Alpert, P., Neeman, B.U., & Shay-El, Y., (1990). Climatological analysis of Mediterranean cyclones using ECMWF data. *Tellus*, *42A*, 65–77.

Appenzeller, C., & Davies, H. C. (1992). Structure of stratospheric intrusione into the troposphere. *Nature*, *358*, 570–572.

Appenzeller, A, Davies, C. H., & Norton,W. A. (1996). Fragmentation of stratospheric intrusions. *J. Geophys. Res.*, *101(D)*, 1435–1456.

Arya, P. (2001). *Introduction to micrometeorology* (2nd edition). San Diego: Academic Press.

Buzzi, A., Tartaglione, N., & Malguzzi, P. (1998). Numerical simulations of the 1994 Piedmont flood: role of orography and moist processes. *Mon. Wea. Rev., 126,* 2369–2383.

Chen, C., & Cotton, W. R. (1983). A One-Dimensional Simulation of the Stratocumulus-Capped Mixed Layer. *Bound.-Lay. Meteorol.*, *25*, 289–321.

Colacino, M., Conte, M., & Piervitali, E. (1997). *Elementi di climatologia della Calabria*. Roma: IFA-CNR.

Fantini, M., & Davolio, S. (2001). Instability of Neutral Eady Waves and Orography. *Jou. Atmos. Sci.*, *58,* 1146–1154.

Federico, S., Bellecci, C., & Colacino, M. (2003a). Quantitative precipitation of the Soverato flood: the role of orography and surface fluxes. *Nuov. Cim. C*, *26 C*, 7-22.

Federico, S., Bellecci, C., & Colacino, M. (2003b). Numerical simulation of Crotone flood: storm evolution. *Nuov. Cim. C, 26 C*, 357-371.

Federico S., E. Avolio, C. Bellecci, A. Lavagnini, & Walko, R. L. (2007): Predictability of intense rain storms in the Central Mediterranean basin: sensitivity to upper-level forcing. *Adv. in Geosciences*, *12*, 5-18.

Federico, S., Avolio, E., Pasqualoni, L., & Bellecci, C. (2008a). Atmospheric patterns for heavy rain events in Calabria. *Nat. Hazards Earth Syst. Sci.*, *8*, 1173-1186.

Federico, S., Avolio, E., Bellecci, C., Lavagnini, A., Colacino, M., & Walko, R. L. (2008b): Numerical analysis of an intense rainstorm occurred in southern Italy. *Nat. Hazards Earth Syst. Sci.*, *8*, 19-35, doi:10.5194/nhess-8-19-2008.

Federico, S., Avolio, E., Pasqualoni, L., De Leo, L., Sempreviva, A. M., & Bellecci, C. (2009). Preliminary results of a 30-year daily rainfall data base in southern Italy. *Atmos. Research, 94*, 4, 641-651, 2009, DOI: 10.1016/j.atmosres.2009.03.00.

Federico S., Pasqualoni L., De Leo L., & Bellecci, C. (2010). A study of the breeze circulation during summer and fall 2008 in Calabria, Italy. *Atmos. Research, 94,* 1-2, 1-13. doi:10.1016/j.atmosres.2010.02.009.

Fehlmann, R., & Davies, H. C. (1999). Role of salient potential vorticity elements in an event of frontal-wave cyclogenesis. *Quart. J. Roy. Meteor. Soc., 125*, 1801-1824.

Fehlmann, R., Quadri, C. & Davies, H. C. (2000). An Alpine rainstorm: Sensitivity to the mesoscale upper level structure. *Wea. Forecasting, 15*, 4-28.

Fisher, H., Eigenwillig, N., & Muller, H. (1981). Information content of METEOSAT and Nimbus/THIR water vapor channel data: Altitude association of observed phenomena. *J. Appl. Meteor., 20,* 1344–1352.

Homar, V., Ramis, C., & Alonso, S. (2001). A deep cyclone of African origin over the Western Mediterranean: diagnosis and numerical simulation. *Ann. Geophys., 20*, 93-106.

Homar, V., Romero R., Ramis, C., & Alonso, S. (2002). Numerical study of the October 2000 torrential precipitation over eastern Spain: analysis of the synoptic-scale stationarity. *Ann. Geophys., 20*, 2047-2066.

Hoskins, B. J., McIntyre, M. E., & Robertson, A. W. (1985). On the use and significance of isentropic potential vorticity maps. *Quart. J. Roy. Meteor. Soc., 111,* 877–946.

Jansa, A., Genoves, A., Picornell, M.A., Campino, J., Riosalido, R., & Catterero, O. (2001). Western Mediterranean cyclones and heavy rain. Part 2: statistical approach. *Int. Jou. of Climatology, 8*, 43–56.

Krichak, S. O., Alpert, P., & Melina, D. (2004). The role of atmospheric processes associated with hurricane Olga in the December 2001 floods in Israel. *J. Hydrometeor., 5*, 1259-1270.

Krichak, S.O., Alpert, P., & Dayan, M. (2007) A southeastern Mediterranean PV streamer and its role in December 2001 case with torrential rains in Israel. *Nat. Hazards Earth Syst. Sci., 7*, 1-12.

Martin, A., Romero, R., Homar, V., De Luque, A., Alonso, S., Rigo, T., & Llasat, C. (2007). Sensitivities of a flash flood event over Catalonia: a numerial analysis. *Mon. Wea. Rev., 135,* 651-669.

Mass, C. F., Ovens, D., Westrick, K., & Colle, B. A. (2002). Does increasing horizontal resolution produce more skilful forecasts? *Bull. Amer. Meteor. Soc., 83*, 407-430.

Mass, C. F., Albright, M., Ovens, D., & cohautors (2003). Regional Environmental Prediction Over the Pacific Northwest. *Bull. Amer. Meteor. Soc., 84,* 1353-1366.

Mellor, G. & Yamada, T. (1982). Development of a Turbulence Closure Model for Geophysical Fluid Problem. *Reviews of Geophysics and Space Physics, 20*, 851–875.

Menzel, W. P., Holt, F. C., Schmit, T. J., Aune, R. M., Schreiner, A. J., Wade, G. S., & Gray, D. G. (1998). Application of GOES-8/9 soundings to weather forecasting and nowcasting *Bull. Amer. Meteor. Soc., 79*, 2059–2077.

Molinari, J. & Corsetti, T. (1985). Incorporation of cloud-scale and mesoscale down-drafts into a cumulus parametrization: results of one and three-dimensional integrations. *Mon. Wea. Rev., 113*, 485-501, 1985.

Molteni, F., Buizza, R., Marsigli, C., Montani, A., Nerozzi, F., & Paccagnella, T. (2001). A strategy for high-resolution ensemble prediction, Part I: definition of representative members and global model experiments. *Quart. J. Roy. Meteor. Soc., 127,* 2069–2094.

Moscatello, A., Miglietta, M. M., & Rotunno, R. (2008). Numerical Analysis of a Mediterranean "Hurricane" over Southeastern Italy. *Mon. Wea. Rev., 136*, 4373-4397.

Pastor, F., Estrela, M.J., Peñarrocha, D., & Millán, M.M. (2001). Torrential Rains on the Spanish Mediterranean Coast: Modeling the Effects of the Sea Surface Temperature. *J. Appl. Met., 40*, 1180–1195.

Pielke, R. A. (2002). *Mesoscale Meteorological Modeling.* San Diego: Academic Press.

Ramond, D., Corbin, H., Desbois, M., Szejwach, G., & Waldteufel, P. (1981). The dynamics of polar jet streams as depicted by the METEOSAT WV channel radiance field. *Mon. Weather Rev., 109*, 2164–2176.

Romem, M., Zi, B., Saaroni, H. (2007). Scenarios in the development of Mediterranean Basin. *Advances in Geosci., 12*, 59–65.

Romero, R., Ramis, C., Alonso, S., Doswell III, C.A., & Stensrud, D.J. (1998). Mesoscale Model Simulations of Three Heavy Precipitation Events in the Western Mediterranean Region. *Mon. Wea. Rev., 126*, 1859-1881.

Romero, R., Martín, A., Homar, V., Alonso, S., & Ramis, C. (2005). Predictability of prototype flash flood events in the Western Mediterranean under uncertainties of the precursor upper-level disturbance: The HYDROPTIMET case study. *Nat. Hazards Earth Syst. Sci., 5*, 505-525.

Rotunno, R., Ferretti, R., 2001. Mechanisms of intense Alpine rainfall. Journal of Atmospheric Sciences 58, 1732–1749.

Santurette, P., and Georgiev, C. G.: *Weather Analysis and Forecasting.* Elsevier Academic Press. 179 pp, 2005.

Siccardi, F. (1996). Rainstorm hazards and related disasters in the western Mediterranean region. *Remote Sens. Rev., 14*, 5-21.

Smagorinsky, J. (1963): General circulation experiments with the primitive equations. Part I, The basic experiment. *Mon. Wea. Rev., 91*, 99-164.

Stein, U., & Alpert, P. (1993). Factor Separation in Numerical Simulations. *J. Atmos. Sci., 50*, 2107–2115, 1993.

Trigo, I.F., Davies, T.D., & Grant, R.B. (1999). Objective climatology of cyclones in the Mediterranean region. *Jou. of Climate*, 12, 1685–1686.

Trigo, I.F., Grant, R.B., & Davies, T.D., (2002). Climatology of cyclogenesis mechanisms in the Mediterranean. *Mon. Wea. Rev., 130*, 549–569.

Walko, R. L., Cotton, W. R., Meyers, M. P., & Harrington, J. Y. (1995). New RAMS cloud microphysics parameterization part I: the single-moment scheme. *Atmos. Research, 38*, 29-62.

Walko, R. L., Band, L. E., Baron, J., Kittel, T. G., Lammers, R., Lee, T. J., Ojima, D., Pielke, R. A. Sr., Taylor, C., Tague, C., Tremback, C. J., & Vidale, P. L. (2000). Coupled Atmosphere-Biosphere-Hydrology Models for environmental prediction- *J. Appl. Met.,* 39, 931-944.

In: Rainfall Forecasting
Editor: T.S.W. Wong, pp. 199-232
ISBN: 978-1-61942-134-9

Chapter 5

PREDICTION OF TROPICAL CYCLONE RAINFALL POTENTIAL IN TAIWAN MOUNTAINOUS AREAS

Gin-Rong Liu[1], Tsung-Hua Kuo[1], Wann-Jin Chen[2] and Tang-Huang Lin[1]

[1] Center for Space and Remote Sensing Research, National Central University, No. 300, Chungda Rd., Chung-Li, Taiwan 320001.

[2] Department of Electric Engineering, Ta Hwa Institute of Technology, No.1, Ta Hwa Rd., Chiung-Lin, Hsin-Chu County, Taiwan 30740.

ABSTRACT

Estimation of the tropical cyclone rainfall potential can be performed by adding together the respective satellite microwave derived rainfall values. However, this technique is essentially based upon a very simple cyclone shift-motion assumption and the cyclone track prediction results. Such assumptions are too simple to correspond well with the actual rainfall conditions, especially after landfall. The absence of taking into account the typhoon rainband rotation could induce significant rainfall location inconsistencies. Even more serious errors may occur if the coupling interactions with the orographic effects produced from mountainous areas are not considered.

In light of these problems, this chapter presents a method in constructing an improved tropical rainfall potential (I-TRaP) technique for mountainous regions by factoring in additional vital information. Using the original rainfall potential method as the basis, the typhoons' rotation speeds are estimated with geostationary satellite imagery and added to the new model. Furthermore, the changes in the typhoon rainfall patterns and rainfall intensity caused by the orographic effect are also considered together with the historical cases, thereby allowing the predicted rainfall patterns to be redistributed and rescaled to approximate the actual rainfall as close as possible.

Typhoon cases which made landfall in the hilly and mountainous regions of Taiwan were retrieved with satellite observations and tested to demonstrate the technique's performance. Results showed that the more detailed considerations rendered much better predictions of the typhoon rainfall potentials, both in terms of the rainfall locations and rainfall accumulation of landfalling typhoons.

Keywords: Central Mountain Range, Coupling effect, I-TRaP, Monsoon, Orographic effect, Rainband, Rainfall potential, remote sensing, satellite, seasonal wind, TRaP, Tropical cyclone, Typhoon.

1. INTRODUCTION

For many geographic regions such as Taiwan, the rainfall from typhoons or hurricanes is one of its key water supply sources. But from another perspective, such tropical cyclones (TCs) may also bring forth unexpected heavy rainfall, thereby causing flash floods, mudslides or other related disasters (Emanuel, 2005). For example, Typhoon Morakot (2009) brought record-breaking rainfall and floods in just three days, greater than any rainfall and flood that occurred in Taiwan over the previous 50 years. It killed more than 670 people (about 500 were killed by a single huge mudslide triggered by the heavy rainfall). The economic losses incurred by Typhoon Morakot were the most severe among the many typhoons that have hit Taiwan. Rainfall records showed that the mountainous areas of southern Taiwan recorded a 72h total precipitation reaching nearly 3000 mm. For many TC cases, flash floods are still the most devastating nature disaster when compared to strong winds or tidal waves, especially for regions such as Taiwan, which owns a steep and high topography (Wu et al., 2002; Zhang et al., 2010). As Typhoon Morakot was only given a category-2 TC status before it made landfall, most people did not anticipate the extreme heavy rainfall. One of the key reasons behind this extreme event was attributed to the interaction between Typhoon Morakot's moisture circulation and Taiwan's Central Mountain Range (CMR). In addition, observations showed that the subsequent southwesterly wind associated with Typhoon Morakot's outer circulation provided a constant source of moisture, which resulted in continuous heavy rainfall over southern Taiwan. The rainfall induced damage clearly illustrates the potential destructive power of a TC. To reduce the damage caused by a TC, it is imperative to develop a method which can forecast the TC rainfall potential accurately. This is particularly important for landfalling cases over hilly and mountainous regions, due to higher flood and mudslide risks caused by rainfall intensification. Through such a TC rainfall prediction model, it could provide a better opportunity and additional precious time to set up more effective disaster mitigation and water resource management plans.

Estimating the rainfall potential of a landfalling TC has been a major operational challenge to forecasters for decades. It can become very difficult when the interaction between the TC system and the inland rugged terrains is strong, which unfortunately is usually the case for Taiwan. Over two-thirds of Taiwan is covered by mountainous terrain where many reach as high as 3000 meters. Thus, the landfalling TC system definitely interacts with Taiwan's complex mesoscale topography, especially the CMR. In addition, the interaction between the CMR, the storm's outer circulation and the southwesterly/northeasterly winds around Taiwan is another factor that further complicates the rainfall potential estimation.

To provide a more accurate accumulated rainfall amount prediction, more comprehensive estimation methods are needed. Although many forecasters rely heavily on numerical weather prediction (NWP) models, they are unable to predict short-term rainfall amounts and patterns very well, due to their various limitations (Ferraro et al., 2005). In fact, it is very challenging to employ any NWP model in predicting cases such as Typhoon Morakot. As mentioned previously, the interactions between the CMR and TC circulations, along with the coupling effect of the seasonal winds are extremely complicated. Precise modeling of the complex interactions and mechanisms via dynamic operational models is nearly impossible. Moreover, the models need to possess sufficient spatial resolution to resolve the mesoscale CMR problem.

Hence, for weather centers or disaster mitigation agencies, a more simple and straightforward method needs to be developed in providing a fast TC rainfall potential prediction, especially for rugged terrain. Employment of satellite observations is a viable approach for such situations. For example, Kidder et al. (2005) used various satellite microwave data, such as the Special Sensor Microwave/Imager (SSM/I), Tropical Rainfall Measuring Mission

(TRMM) Microwave Imager (TMI) and Advanced Microwave Sounding Unit (AMSU) in predicting the 24-h tropical rainfall potential. The results showed how satellite data performed better than operational mesoscale models.

Today no tropical cyclone can evade being detected by satellites. Naturally, not only do forecasters want to know their exact locations; they also seek to understand how the storm behaves, or even predict how it will evolve in the future. Therefore, satellite-monitoring of TCs has always been one of the main tasks for researchers. The well-known T-number technique proposed by Dvorak (1975) is the first important successful satellite application of TC monitoring. The method basically classifies TCs into different T-number categories by their cloud patterns and characteristics. This approach made possible for the estimation of the TC intensity (Dvorak, 1984). Considered a simple but powerful application, it inspired many successive researchers to develop other satellite-based TC monitoring methods. In addition, the continued improvements in the satellite channels and resolutions have also been an important factor behind the satellite's growing versatility. For example, utilization of satellite infrared images in estimating the TC rainfall has already been conducted for years. But due to the technical developments of satellite microwave remote sensing in recent decades, the rainfall retrieval accuracy has greatly improved (Alliss et al., 1992; Farraro, 1997; Kidder et al., 2000; Lonfat et al., 2004), thanks to the better application ability in TC rainfall monitoring and potential prediction.

Development of calculating the tropical cyclone rainfall potential has a long history. In the 1960s, Riehl and Malkus (1961) already pointed out that rainfall decreased exponentially with the distance from the TC center. Following these findings, the maximum 24-h rainfall as a function of storm translation speed was used to establish the first operational TC rainfall estimation model. Many successive models were developed in later decades, mainly for assessing potential flooding risks (Pfost, 2000). These earlier operational applications were conducted with satellite infrared images. Following the launch of cloud-penetrating microwave observations, the TC rainfall potential could be estimated more accurately. In the late 1990s, the technique was greatly improved in several ways for higher automation and flexibility. The most important was the tropical rainfall potential (TRaP) technique proposed by Kidder et al. (2001, 2005) for operational applications. However, the basic assumptions in the method were still too simple to approximate the actual TC rainfall behavior. The assumptions are summarized as follows: (1) The forecast TC track is correct. (2) The satellite-derived rainfall rates are correct. (3) The spatial pattern of rainfall rates relative to the

TC center does not change in either coverage or magnitude, and thus moves with the TC center along the predicted track (Kidder et al., 2005). It is very clear that the assumptions neglect many factors, such as (1) the rotation motions of a TCs' spiral rainbands, (2) the intensity change, and (3) the outside environmental influence. An example of this influence is the interaction between the TC system and the terrain, especially when it is hilly and mountainous. It is obvious that the orographic effect is significant and must be taken into account if a more reasonable rainfall potential pattern is to be retrieved (Ramsay and Leslie, 2008).

As such, rainfall potential schemes mostly adopted very simple assumptions, and generally did not consider the orographic effect on the rainfall. The derived rainfall potential was thus inaccurate, especially for rugged terrain. In order to develop a more practical TC rainfall potential model, which included the orographic effect, a new comprehensive TC rainfall potential model is introduced in this chapter. The step-by-step improvements will be shown to demonstrate its practicality.

The original TRaP technique only considered the shift-motion of TCs. However, it is well-known that a TC is essentially a large rotating rainfall system, where the rainfall distribution pattern is generally not in a steady-state. The revolving features can be recognized from the spiral rainbands observed by consecutive satellite visible and infrared images (Liu et al., 2008). Although some polar-orbiting images can be used to estimate the rotation speeds (Wimmers, 2007), their observations are not updated frequently enough to represent the actual rainband evolution. Therefore, a better approach is to take into account the feature of the rainband rotation from hourly geostationary satellite images. The TC rainfall distribution that corresponds to its rainband rotation could then be taken into account. Liu et al. (2008) estimated the TC angular rotation speed from two consecutive geostationary weather satellite images by using the maximum cross-correlation method. It was applied to improve the original TRaP technique (Kidder et. al, 2005), which was subsequently used to estimate the TC rainfall accumulation via extrapolating the retrieved rain rates by following a forecast track.

Yet, even by including the influence of the rainband rotations, it is still unlikely to estimate the tropical cyclone rainfall potential accurately if the orographic effect is excluded, especially for mountainous areas. For Taiwan, the interactions between TC circulations and the rugged terrain is rather complex (Yen and Elsberry, 1993, Lin et al., 2001). Generally, this cannot be easily numerically simplified and handled by simulation models. For operational applications, many researchers have tried to establish more straightforward approximations. For example, Wang et al. (1986) used typhoon locations over the ocean and the future corresponding typhoon rainfall over land to find the rainfall potential estimation relationship. The method provided a rough approximation of the mountainous terrain influence and TC tracks. However, their method considered the TC locations as the only influencing factor and neglected other variables, such as typhoon rainfall intensity, size and circulations and so on. Cheung et al. (2008) used the climatology-persistence (CLIPER) method to estimate the rainfall potential. The climatology part owned a similar concept as the method proposed by Wang et al. (1986) where the historical statistical rainfall distributions were also considered. Their result showed that the correlation coefficients with the station measurements were roughly $0.6 \sim 0.7$ for 6-h accumulated rainfall rates, revealing the historical long-term rainfall pattern could be used as an correction factor for the orographic effect on rainfall patterns. In other words, TCs approaching Taiwan with a similar track seemed to render a similar rainfall pattern over the island. Therefore, based upon the original rainfall potential technique, Liu (2009) classified typhoons around Taiwan into several track sections to investigate the orographic effect on the tropical rainfall potential. The result pointed out that the terrain effect should be considered to improve the original tropical rainfall potential technique in delineating a typhoon's rainfall potential, especially for landfalling typhoons over rugged terrain.

This chapter mainly details the improvements to the original tropical rainfall potential technique. The data involved in the technique is introduced in Sections 2 and 3. The step-by-step improvements are also described. The performance of the improved tropical rainfall potential for mountainous areas is illustrated in Section 4, followed by a conclusion in Section 5.

2. Dataset Description

The Special Sensor Microwave Imager (SSM/I) data from the Defense Meteorological Satellite Program (DMSP) sun-synchronous satellites, along with geostationary satellite infrared data were used to predict the TC rainfall potential. The SSM/Is are passive microwave radiometers scanning in the dual-polarized 19.4, 37.0 and 85.5GHz channels and vertical-polarized 22.2GHz channels (Hollinger et al., 1987). The data were used to estimate a TC's rainfall distribution. The rainfall data is subsequently summed up for retrieving the accumulated rainfall distribution. The geostationary satellite data included the Geostationary Meteorological Satellite 5 (GMS-5), Geostationary Operational Environmental Satellite 9 (GOES-9) and Multifunctional Transport Satellite-1R (MTSAT-1R). The satellites provided meteorological data over the Asia-Pacific region. The GMS-5 was located at 140°E longitude and 35,800 km above the equator. In May 2003, its service lifetime ended, and was temporarily replaced by GOES-9, which was located at longitude 155°E. In June 2005, the MTSAT-1R officially succeeded the GMS-5. The hourly infrared images scanned by these geostationary satellites provide an opportunity to estimate the rotational speed of the tropical storms' spiral rainbands. To ensure image consistency, all the images were remapped to an approximate 2.7 km pixel resolution by using the Lambertian conformal projection.

In addition to the satellite data, the 6-h best track data from the US Navy's Joint Typhoon Warning Center (JTWC) were used in positioning a typhoon's center and delineating its track. Rainfall rates estimated by SSM/I were compared with the hourly rain gauge data. The data were collected by the automatic rain gauge network covering the entire Taiwan island (Figure 1) established by Taiwan's Central Weather Bureau. The density of the rain gauge stations over the island has significantly increased in recent years. This is a great help to the detailed analysis of the island's heavy rainfall. In this chapter, typhoons during the period 1997 to 2009 were collected.

3. The Improved TRaP (I-TRaP) Technique

The original TRaP technique starts with a rainfall estimation from satellite-based observations along with the observed storm tracks. The position of the storm center is obtained every 15 minutes during the following 24-h period by interpolation. Afterwards, the TC's rainfall potential can be obtained by summing up all the single rainfall estimations. The TRaP method is capable of quickly predicting the next 24-h rainfall distributions before TC landfall, which is very helpful if flood warning and evacuation orders need to be issued.

However, the technique is based upon very simple assumptions. It only considers the advection of rainfall patterns. Therefore, the improved tropical rainfall potential technique (referred as I-TRaP) proposed in this chapter considers additional factors regarding a TC's rainfall pattern. This includes its rainband rotations, and the orographic effect. A redistribution and rescale procedure is conducted to correct the original TRaP result and produce a new and more precise rainfall pattern.

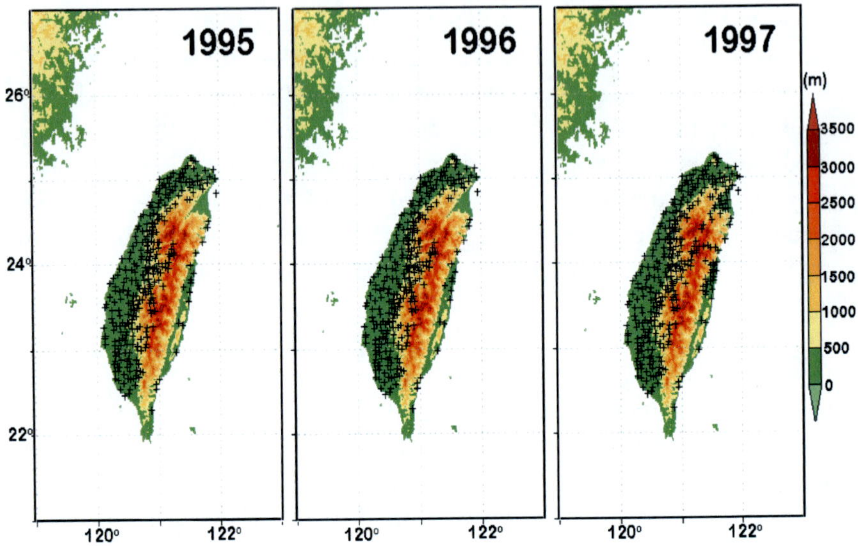

Figure 1. The automatic rain gauge network in Taiwan. The density of the rain gauge stations has significantly increased in recent years. Meanwhile, the north-south orientation of the CMR is clearly shown. The unit of the color bar is in meters.

3.1. Rainfall Estimation with Satellite Microwave Data

Due to the cloud-penetrating ability, which provides better accuracy in rain rate estimation (Chiu et al., 1990; Ferraro, 1997), microwave data such as the SSM/I data, was a more suitable choice over infrared images for TC rainfall potential computation. The primary algorithm used in this chapter is an 85 GHz scattering-based technique over both land and ocean; combined with the 19/37 GHz to consider the emission effect over the ocean. In the rainfall estimation, a pair of scattering indexes (SI) are respectively involved for the land and ocean, (Ferraro et al., 1994). Thus, the rainfall rates over the land and ocean can be obtained separately as,

$$RR = 0.00513 SI_L^{1.9468} \quad (1a)$$

$$RR = 0.00188 SI_W^{2.0343} \qquad (1b)$$

and where

$$SI_L = [451.9 - 0.44 * TB_{19V} - 1.775 * TB_{22V} + 0.00575 * TB_{22V}^2] - TB_{85V} \qquad (2a)$$

$$SI_W = [-174.4 + 0.72 * TB_{19V} + 2.439 * TB_{22V} - 0.00504 * TB_{22V}^2] - TB_{85V} \qquad (2b)$$

where RR is in mmh^{-1}. The subscript L and W indicate the land and ocean surfaces, respectively. TB is the brightness temperature and the subscript numbers of 19, 22 and 85 denote the SSM/I channels, where the unit is in Kelvin. Figure 2 shows a typhoon rainfall estimation example. The heavy rainfall areas, especially around the eye wall areas, are clearly seen. Studies have found that a SI value greater than 10K is a suitable global rain threshold (Grody, 1991). In addition, the RR values above 35 mmh^{-1} were set to 35 mmh^{-1}, an indication that the maximum rain rate was retrievable with the SSM/I 85 GHz data. On the other hand, the minimum retrievable rain rates are 0.45 and 0.20 mmh^{-1} for land and ocean, respectively, when the 10K minimum threshold for the SI value was used. As an ice-scattering signature may be detected in mid and high-latitude rain systems or tropical warm rain system over the ocean, rain may not be detected by the 10K rain threshold. For such situations, a liquid water emission technique with 19 and 37 GHz channels is employed (Weng and Grody, 1994),

$$Q_{19} = -2.70[\ln(290 - TB_{19V}) - 2.84 - 0.40\ln(290 - TB_{22V})] \qquad (3a)$$

$$Q_{37} = -1.15[\ln(290 - TB_{37V}) - 2.99 - 0.32\ln(290 - TB_{22V})] \qquad (3b)$$

where Q_{19} and Q_{37} are the liquid water estimated with the 19 and 37 GHz channels, respectively. Afterwards, the rain rates can be obtained if either Q_{19} > 0.60 mm or if Q_{37} >0.20 (Ferraro and Marks, 1995),

$$RR = 0.001707(100Q)^{1.7359} \tag{4}$$

where the Q value is either the Q_{19} or Q_{37} value. The maximum and minimum retrievable rain rates are 0.30 and 35 mmh^{-1}, respectively.

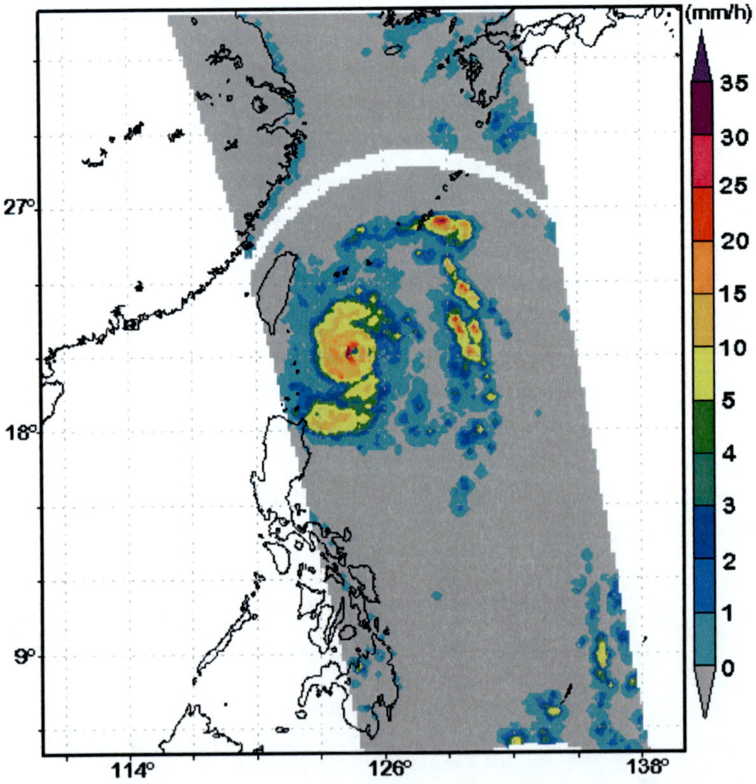

Figure 2. An example of the rainfall distribution estimated by SSM/I data (0934 UTC 27 Sep 2008) and the Farraro's algorithm. Unit in mmh^{-1}. The figure denotes Typhoon Jangmi, which brought forth heavy rainfall and flash floods to northeastern Taiwan.

3.2. TC Rainband Rotation

The spiral rainband of a TC rotating in a counterclockwise manner can be seen via consecutive satellite images. The rotation motion can also be witnessed by ground-based radars (Senn and Hiser, 1959). The original TRaP technique assumes the rainbands of tropical cyclones do not change during the next 24 hours. However, this is not how a TC actually behaves. The study on Hurricane Andrews (1992) conducted by Gall et al., (1998) found that the small scale radar reflectivity of TCs moved in a near tangent direction, and the rotation speeds vary during the different typhoon lifespan stages. However, considering the temporal resolution of satellite observations, it is assumed that the spiral rainband owns a constant rotation speed between two consecutive images in this chapter.

Although several SSM/I sensors are equipped on the DMSP series satellites, the shortest time interval between two consecutive SSM/I observations in low latitude regions is below half an hour. However, sometimes the longest is more than three hours, making it impossible to provide the hourly rotational speeds of a TC with SSM/I data. Consequently, Earth synchronous satellite hourly images are used in this chapter to estimate the rainband rotation speeds. Estimation of the rotation speeds are conducted from satellite image sequences by employing the maximum cross-correlation method, which are delineated as follows: (1) a circle with a 150 pixel (~400 km) radius is cut out of the TC's center in every Lambertian image. (2) The correlation coefficient of the image with the subsequent one is calculated by comparing the rotation speed for each degree, ranging from 0 to 30 degrees. (3) The rotation speed that has the maximum correlation coefficient and minimum RMSE is chosen to represent the storm's rotation speed. This method treats tropical storms as a rigid body, where every point of the typhoon has the same angular rotation speed. An example with Typhoon Dujuan (2003) is shown in Figure 3. Two consecutive hourly images observed by the GOES-9 geostationary satellite are used to calculate different angles' correlation coefficients and RMSE of the two images. The ones with the largest correlation coefficient and lowest RMSE are selected as the storm rotation speed. For the Typhoon Dujuan case, the rotation angle is shown to be 7 degrees. Thus, the rotation speed of the typhoon is set to 7 degree h^{-1} during that time.

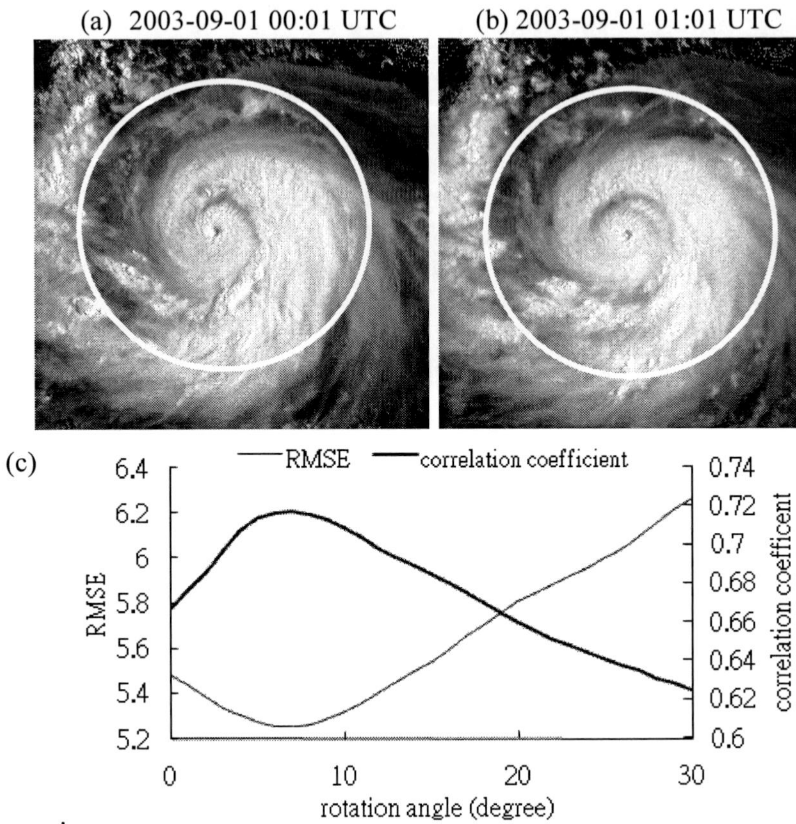

Figure 3. (a) GOES-9 visible satellite image observed at 0001 UTC 1 Sep 2003, (b) the next hourly image, (c) correlation coefficient (bold line) and RMSE (thin line) for rotation angles. The maximum correlation coefficient and the minimum RMSE is found at the 7 degree rotation.

3.4. Orographic Effect on Rainfall

The island of Taiwan possesses a complex mesoscale topography. Over 250 mountains in the CMR exceed a height of 3000 m, and the highest mountain - Jade Mountain has a peak of nearly 4000 meters (3952 m to be exact), which is the highest peak in the Northeastern Asia region. Yet, the island merely has a dimension of roughly 300 km by 100 km. This steepness and roughness is considered significant and unique. As Taiwan is situated in one of the frequent typhoon paths, the interaction between the CMR and typhoons is undoubtedly significant. For example, it has been shown that the

CMR can cause typhoon tracks to deviate (Brand and Blelloch, 1974), or significantly reduce the intensity when typhoons make landfall. Most importantly, the orographic effect produces a complex influence on the typhoon precipitation. The topographic-enhanced moisture flux is difficult to parameterize and model accurately. In other words, the orographic effect makes the rainfall potential difficult to estimate. A trial by Wang et al. (1986) provided an alternate approximation to solve Taiwan's CMR orographic effect on the typhoon precipitation estimation. Historical typhoon rainfall data is used, which mainly include typhoon locations before landfall and the corresponding precipitation over land when the storm is near Taiwan or has made landfall, to construct a location-precipitation prediction. The approximation is reasonable because typhoons in the northern hemisphere always rotate in a counterclockwise direction. A typhoon's circulation generally has similar interactions with the CMR when it is within a certain geographical location that is near or over Taiwan. For example, when a typhoon approaches northern Taiwan, the western side of the CMR is upwind, while on the other side, it is downwind. The opposite occurs when a typhoon hits southern Taiwan. The upwind and downwind areas will receive different rainfall activity. Hence, the historical rainfall over Taiwan and the corresponding typhoon locations intrinsically contain the orographic effect information. However, such a method merely provides a statistical aspect and does not consider the individual typhoon's rainfall contribution. Based on such a concept and by further incorporating the satellite TRaP estimation, a refinement can be made to compute a TC's rainfall potential, which simultaneously contains statistical and actual individual typhoon rainfall information. Hence, it is a promising approach to retrieve the orographic-effect rainfall reasonably even if the complex physical interaction is not specifically dealt with.

Technically speaking, the original TRaP-derived rainfall potential contains no orographic effect, but the station-measured rainfall pattern could be treated as the result that is actually affected by the orographic effect. Therefore, the long-term means of these two sets of rainfall patterns could help establish a correction relationship to take the orographic effect into account. For example, Figure 4 illustrates the historical typhoon tracks between 1997 to 2008. When the tracks move into the big dark circle, they share a similar geographical location. Figure 5(a) is the mean of the rain gauge measured ground truth for all typhoon tracks moving into the big dark circle, which shows a significant orographic-affected pattern. On the other hand, Figure 5(b) is the mean rainfall amount estimated by the original TRaP technique for all typhoon tracks moving into the big dark circle. It is apparent that no orographic pattern could be found in the rainfall pattern. As mentioned previously, the interactions between the counterclockwise circulations and the CMR induce more rainfall amount over the western (upwind) side. This is the most likely rainfall distribution pattern for a typhoon that moves into the big dark circle areas. Therefore, a redistributing procedure is developed in this chapter.

For a certain typhoon within the big dark circle in Figure 4, its hourly total islandwide rainfall amount (R_T) estimated by the original TRaP technique could be written as

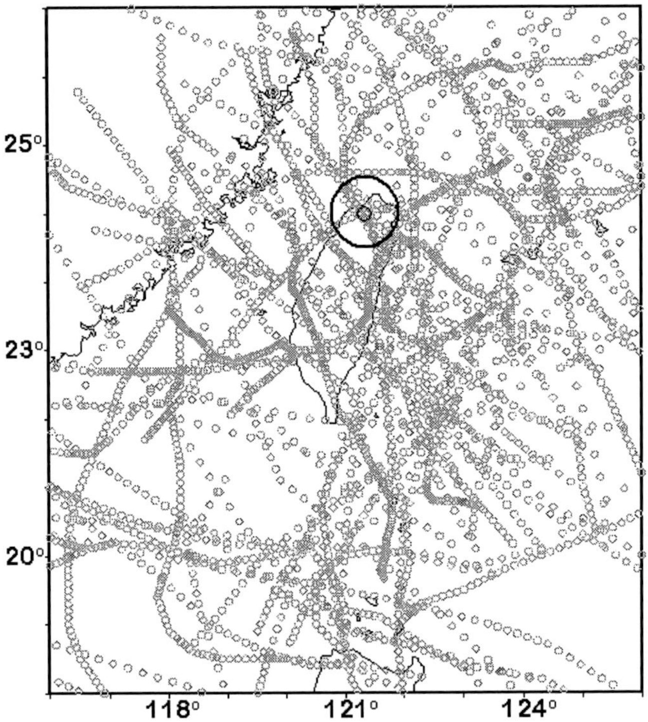

Figure 4. The grey dots represent the hourly position of the respective typhoons that were in the vicinity of Taiwan between 1997 to 2008. The small dark circle denotes the position of a particular typhoon. It also represents the center of the outer dark circle (sampling circle), which has a 1° diameter. For any typhoons that had taken a path crossing this dark circle would be seen as having a similar location and orographic effect with the aforementioned particular typhoon.

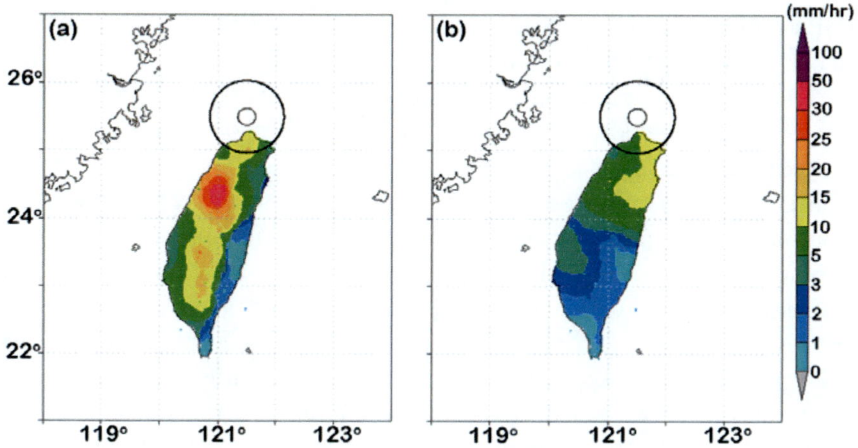

Figure 5. (a) The hourly mean of the rainfall rates for all historical typhoon cases moving into the big dark circle. (b) Same as (a), but from the original TRaP estimation.

$$R_T = \sum_{n=1}^{N} R_{TRaP}(n) \qquad (5)$$

where n is the station ID, N is the total rain gauge number islandwide, and $R_{TRaP}(n)$ is the rainfall amount by the original TRaP for a single rain station. Meanwhile, the corresponding hourly islandwide total rainfall amount (R_S) for historical typhoons observed by stations is

$$R_S = \sum_{n=1}^{N} R_{obs}(n) \qquad (6)$$

where $R_{obs}(n)$ is the observed total rainfall amount for historical typhoons for a single station. The corrected hourly rainfall amount (R_c) for a single rain gauge station could be obtained by

$$R_c = R_T\,(R_{obs}(n) \,/\, R_S) \qquad (7)$$

Basically, it is a normalized redistribution procedure. In this approach, the information of the climatological rainfall pattern is added into each single

TRaP estimation. In other words, the orographic contribution is accounted for and has a new rainfall distribution.

In addition to considering the rainfall pattern redistribution, the difference between the estimated and observed total rainfall amount also needs to be corrected. The difference between the two rainfall amount is obvious. Basically, the bias could be induced by several reasons, such as the rainfall algorithm, the satellite data, the environmental condition and so on. For example, the Farraro algorithm and the 10K SI value is more of a global consideration. Meanwhile, the upwind and downwind, as well as the seasonal wind effects on the rainfall amounts are possible factors. Therefore, four (linear, polynomial, logarithmic, and exponential) equation relationships were used to obtain the corrected factor for the rainfall amount by comparing the TRaP-derived and station-observed rainfall data. The coefficients could be used to reduce the difference of the rainfall potential.

4. Result and Discussion

(a) Rotation of Typhoon Spiral Rainbands

By employing the maximum correlation method and the consecutive hourly images, the rotation speed of a typhoon spiral rainband can be estimated and be further taken into account when calculating the rainfall potential. An example is in Figure 6 for Typhoon Billis (2000). Figure 6(a) shows the rainfall distribution estimated by SSM/I data observed on 0913 UTC 21 August. Figs. 6(b) and 6(c) illustrate the predicted 3-h later rainfall distribution by considering the advection of Typhoon Billis, but a ten degree rainband rotation motion for each hour is further included in Figure 6(c). Figure 6(d) is the observed rainfall distribution from SSM/I on 1209 UTC 21 August, which is three hours later than Figure 6(a). From Figs. 6(b) and 6(d), it can be inferred that the timing of these two images is essentially the same, but the orientations of their eye walls are evidently different. The eye wall in Figure 6(b) is positioned from the northwest to the southeast, but the actual orientation should be from the northeast to the southwest, as observed from the satellite in Figure 6(d). Figure 6(c) shows that when a ten degree rotation estimated by the maximum correlation method is conducted, the result looks more similar with Figure 6(d), especially for the eye wall orientation and outer spiral rainbands.

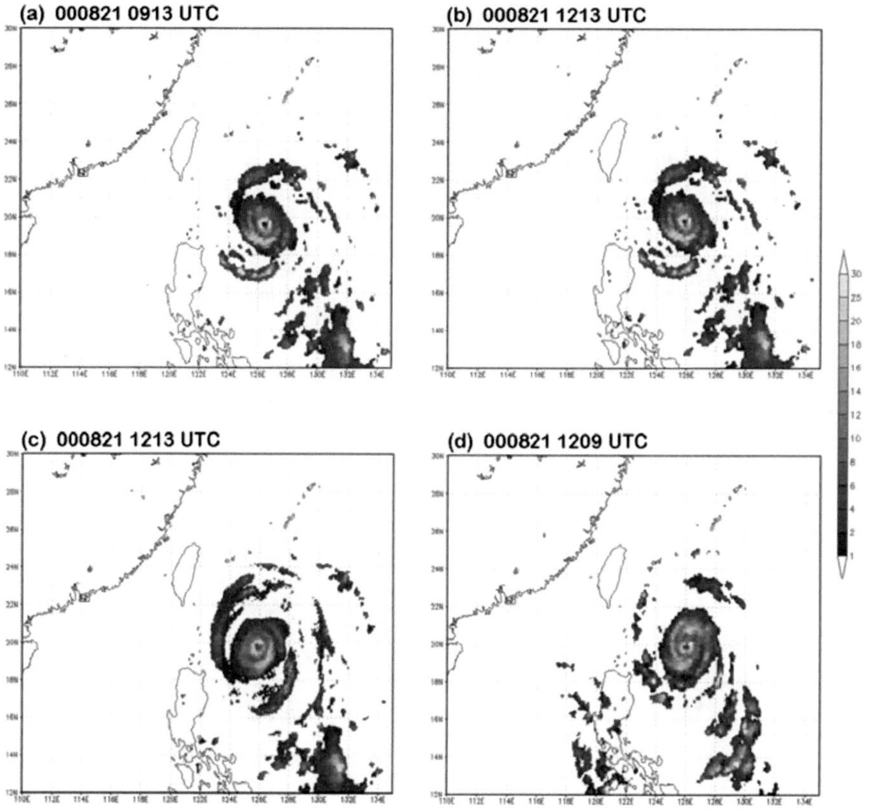

Figure 6. (a) SSM/I-derived rainfall distributions of Typhoon Bilis (2000) at 0913 UTC 21 Aug 2000; (b) Same as (a), but for the possible rainfall distribution at 1213 UTC which is forecasted via the original TRaP technique and use of JTWC's best typhoon track; (c) Same as (b), but rotated by a 10 degree angle; (d) The satellite-observed rainfall distribution at 1209 UTC.

The 6-h rainfall potential can be predicted by advecting and summing up the SSM/I rainfall estimations according to the JTWC best track and satellite-estimated typhoon rotation speed. The correlation coefficients and RMSE comparison between the predicted rainfall and rain gauge data reveal that the accumulated rainfall predictions exhibit obvious improvements over the original TRaP technique when the rotation motions were considered for most typhoon cases.

(b) Relationship between Rainfall Locations and Typhoon Locations

Based upon Figure 5, a relationship appears to exist between the locations of the typhoon centers and rainfall areas. Although this is quite reasonable from a physical science aspect, additional examples are still studied to further confirm the relationship.

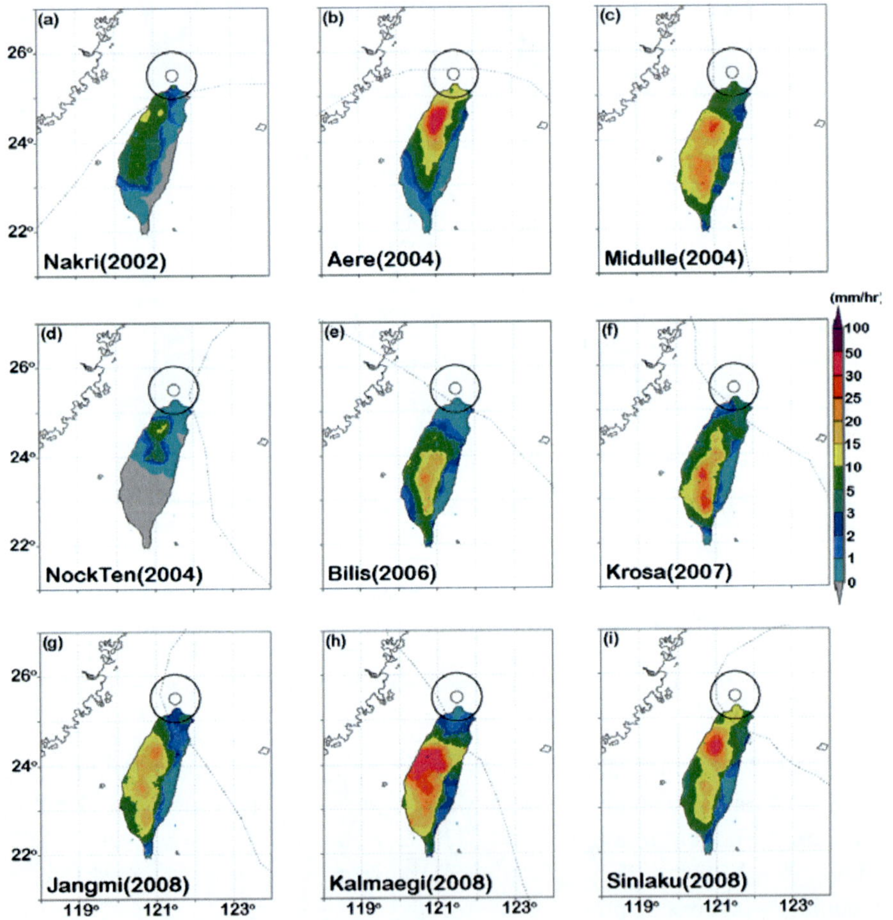

Figure 7. Relationship between the typhoon and rainfall locations. The big dark circle is the sampling circle with 1° diameter centered at (121.5E, 25.5N). The black lines are typhoon tracks. The rainfall distributions over Taiwan are shown in color, unit in mmh^{-1}.

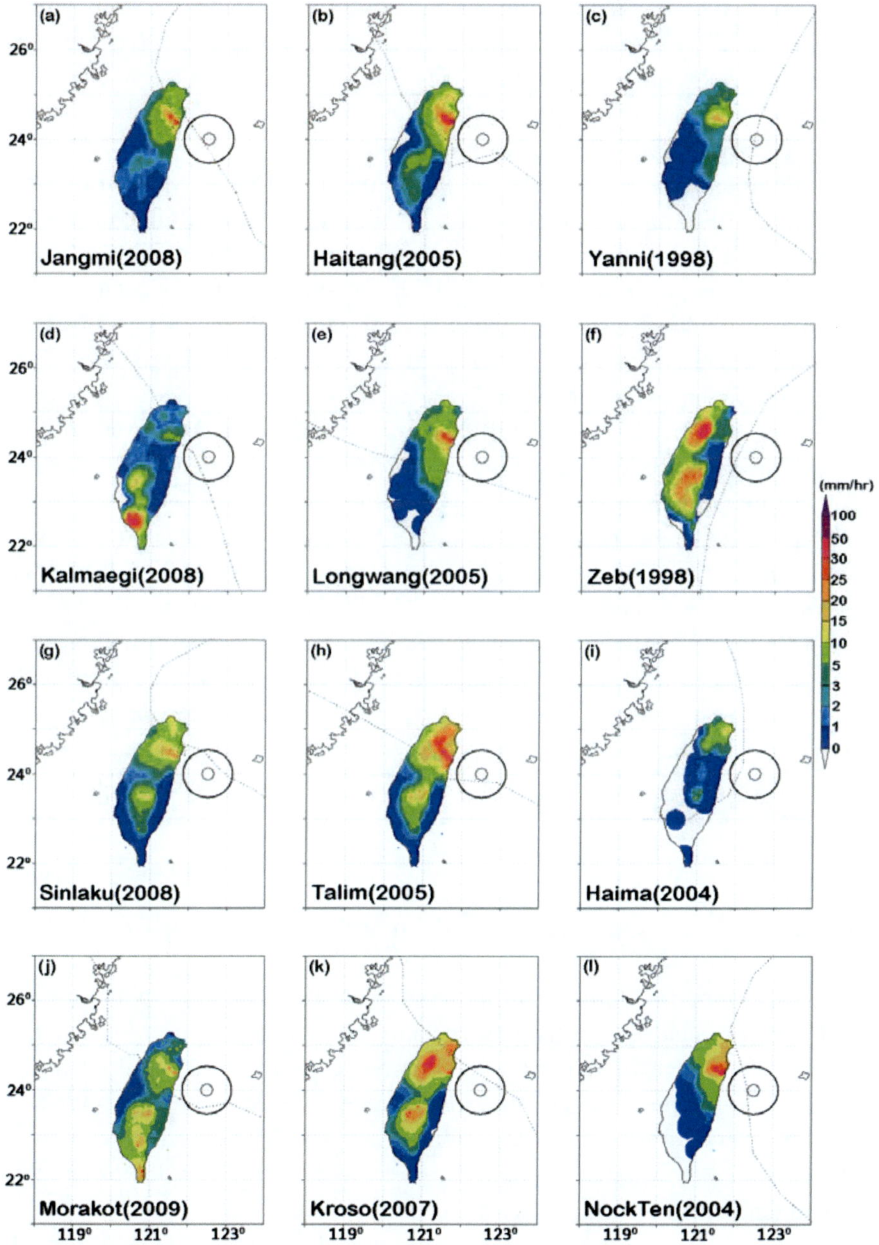

Figure 8. Relationship between the typhoon and rainfall locations. The sampling circle is centered at (122.5E, 24.0N).

When a typhoon moves toward a particular location, a circle with a diameter of 1° is drawn around the location (hereafter referred as "the sampling circle"). Any other typhoons that pass within the sampling circle will be treated as owning a similar location and producing a similar orographic effect on the rainfall distribution. For example, when a typhoon moves toward a location situated at (121.5E, 25.5N), any other historical typhoon that had moved into the sampling circle would be seen to have a similar interaction with the CMR and produce a similar rainfall distribution over Taiwan (Figure 7). In these cases, the rainfall brought by these typhoons mainly occurred over the windward side of western Taiwan, except for the cases in Figs. 7(a) and 7(d), which show a weak rainfall pattern. In addition, nearly all of the historical typhoon cases either brought weak or even no rainfall to the leeward side of eastern Taiwan. The significant relationship in the typhoon location with the rainfall locations under the orographic effect exerted by the CMR is clearly demonstrated.

Meanwhile, another location situated at (122.5E, 24.0N) is analyzed in the same manner. The results show that when a typhoon is located off the eastern Taiwan waters, northeastern Taiwan is most likely to experience the heaviest rainfall (Figure 8). From a physical science perspective, this is also considered reasonable, due to the area's upwind nature. Furthermore, as these particular typhoon cases are already very close to the land areas, the terrain-rainfall activity could induce more precipitation, especially over Yi-Lan County (the red color area in Figs. 8(b), (e) and (l)), which is a plain located to the east of the CMR.

(c) Redistribution and Rescale of Orographic Rainfall

As discussed in Section 3.4, the orographic effect on the typhoon rainfall could be corrected by redistribution and rescale equations via historical typhoon rainfall data. The rainfall amount retrieved from the TRaP technique and rain gauge were used to establish the rescale equation. Four types of equations were tested and it was found that the TRaP technique notably underestimated the rainfall amount. This was likely caused by the orographic-enforcement that the original TRaP could not deal with, or by the Farraro algorithm, which is designed mainly for global applications. However, the orographic effect for each sampling circle could be corrected in a linear relationship,

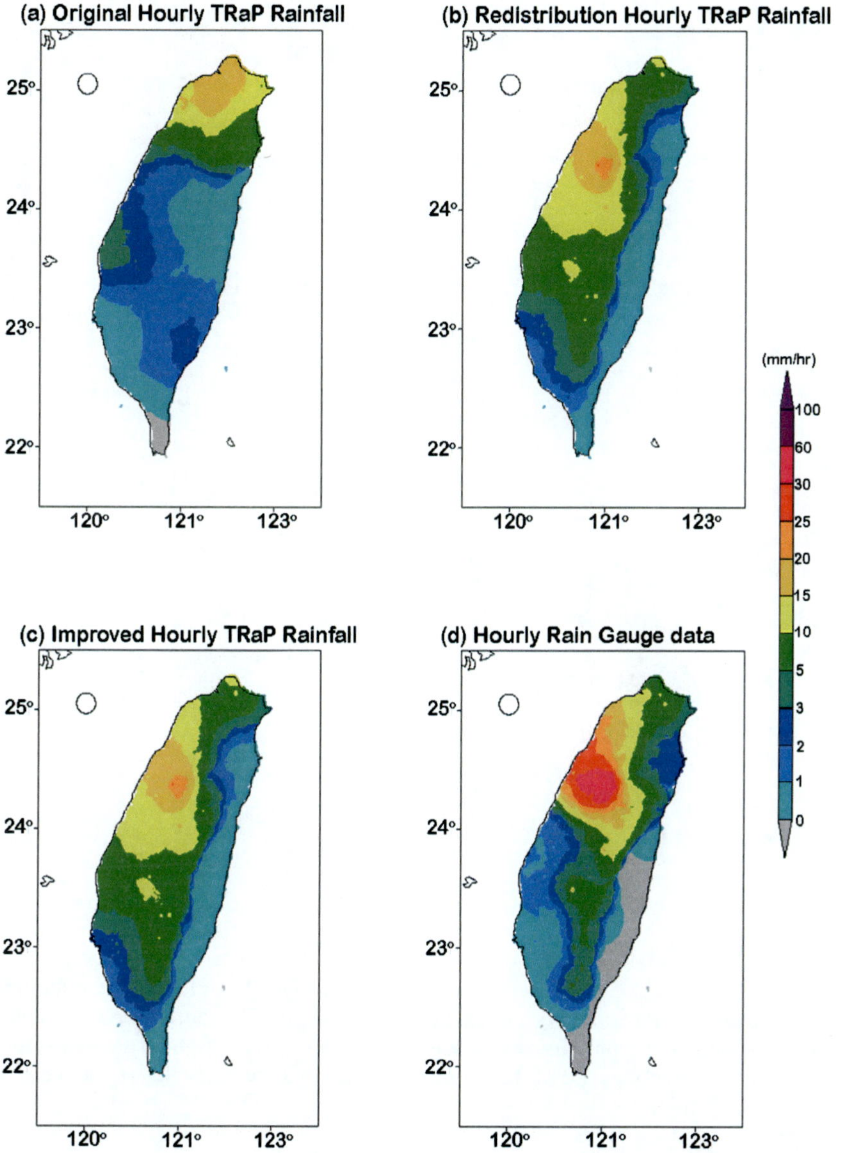

Figure 9. An example of the hourly rainfall distribution (Typhoon Morakot at 1700 UTC, 8 Aug 2009) with orographic correction. (a) TRaP approach where only the rainband rotation is considered. (b) Same as (a) but with redistribution. (c) Same as (a), but with redistributing and rescaling. (d) The rain gauge observed rainfall. The circle indicates the typhoon center.

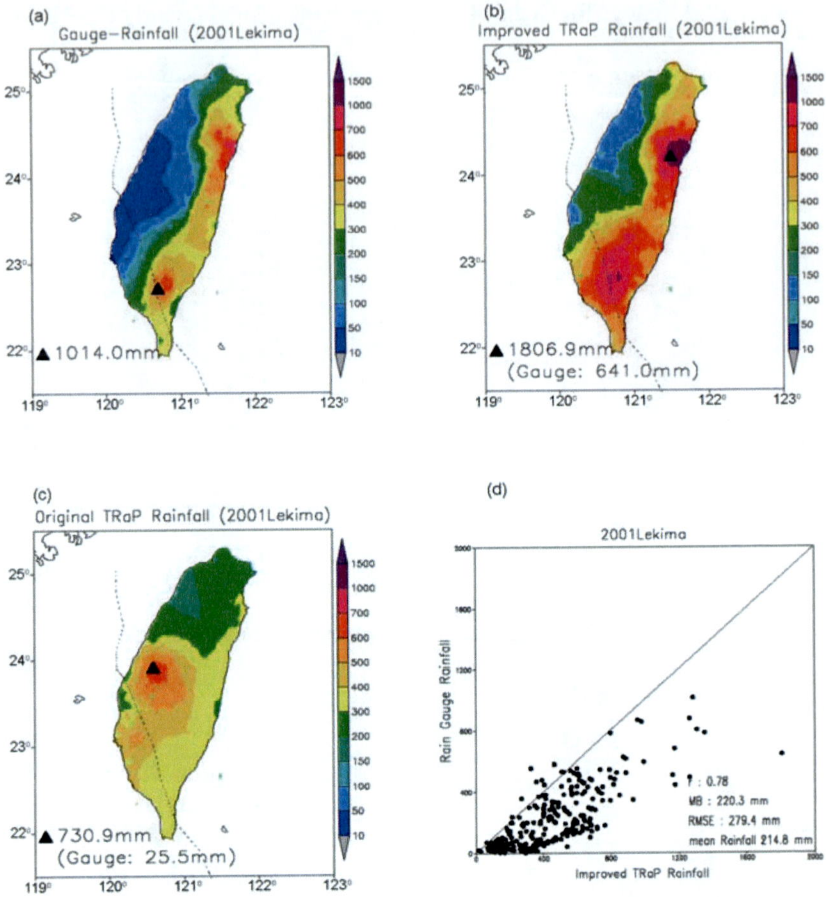

Figure 10. The rainfall potential of Typhoon Lekima (2001). (a) Rain gauge measurement. (b) The I-TRaP result. (c) The original TRaP result. (d) The relationship between the I-TRaP result and rain gauge measurements. The rainfall accumulated time starts when the typhoon track (black line) moves into the framed area and stops once the typhoon moves out of the frame or dissipates as seen for this particular case.

$$R_{cc} = aR_c + b \tag{8}$$

where R_{cc} is the final estimated rainfall amount, R_c is the corrected hourly rainfall amount (see also Section 3.4) and a and b are the rescale coefficients, which can be obtained by comparing the rainfall amount retrieved from the TRaP technique and rain gauge data.

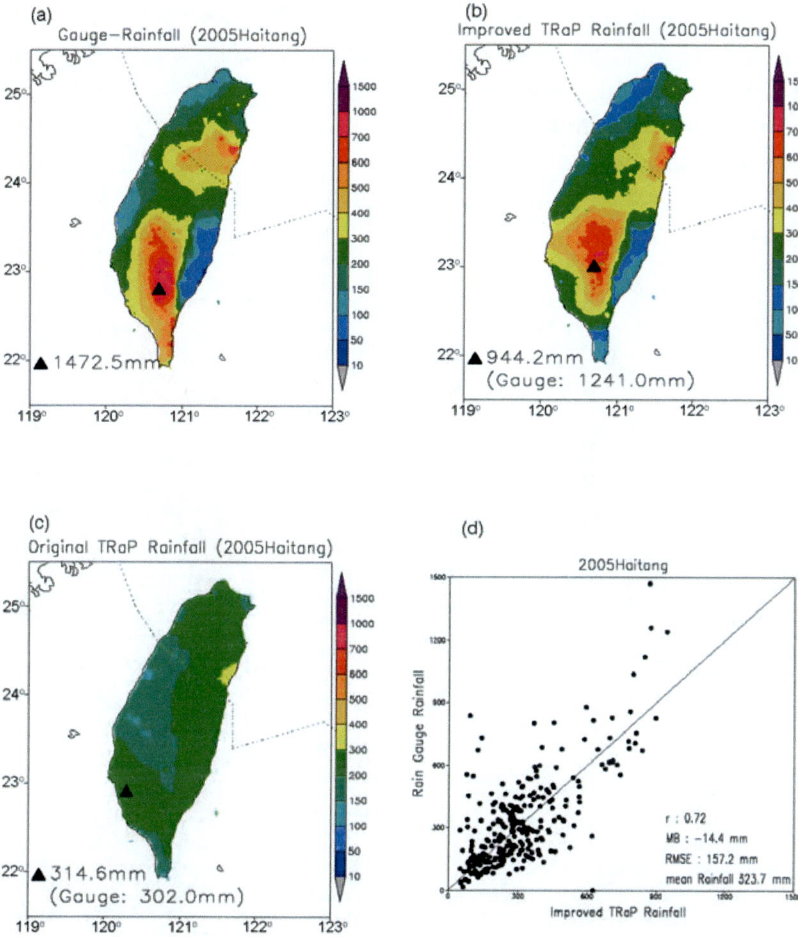

Figure 11. The rainfall potential of Typhoon Haitang (2005).

Figure 9 is an example of hourly rainfall correction by considering the rainband rotation and orographic effect. Comparison between the correction and ground truths shows that the satellite-derived rainfall distribution has a smoother texture. Such differences are often witnessed in the comparisons between satellite and station-measure rainfall data. These two sets of rain rates have an extreme difference in the spatial resolution. Satellite microwave observations always cover a wide area ranging from ten kilometers to several tens of kilometers, but the rain gauges measure the rainfall information on a point-by-point basis, meaning that the former collects the rainfall information from a more macro perspective, while the latter provides a more detailed and localized dataset. In addition, satellites have a rougher temporal resolution when compared to rain gauge measurements. This is another possible cause behind the smoother rainfall pattern in Figure 9(c). Naturally, the rescaling equation should be tested and analyzed further with more historical typhoon cases.

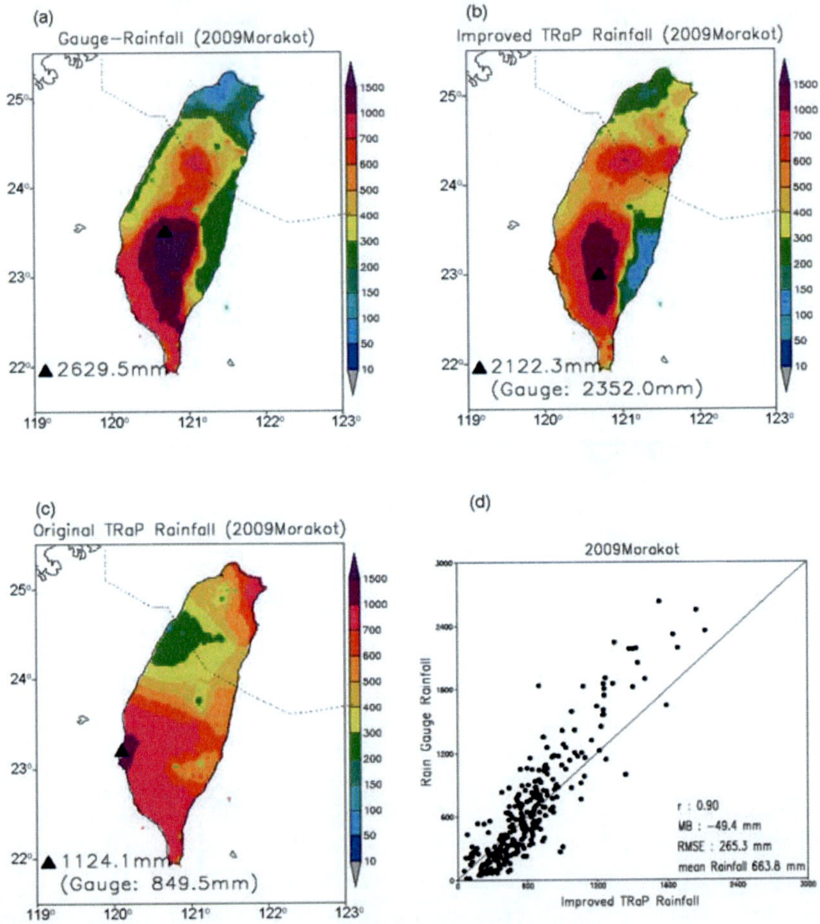

Figure 12. The rainfall potential of Typhoon Morakot (2009).

After establishing the necessary improvements and correction procedures, the 58 non-landing typhoon cases during 2000 to 2008 were tested. Results show that the mean correlation coefficient between the original TRaP result with the rain gauge rainfall rates was merely 0.10 and the RMSE was 120.6mm, an indication that almost no relationship existed between the two. However, the value increased to 0.54 and the RMSE was 99.8mm after the I-TRaP was applied. Aside from the above mentioned cases, a total of 22 landfalling typhoons during the same time period were also investigated. The mean correlation coefficient values were 0.07 and 0.64, and the RMSE values were 196.4 and 130.8 mm for TRaP and I-TRaP respectively. This also demonstrated that I-TRaP gave better predictions than the original TRaP method. Meanwhile, the landfalling cases saw a higher improvement over the non-landfalling cases. Three typhoon landfalling examples, Lekima (2001), Haitang (2005) and Morakot (2009) are shown in Figs. 10-13. Figure 10 is the result of Typhoon Lekima (2001). From the rain gauge data, the rainfall distribution over the island demonstrates an obvious orographic-affected pattern. Eastern Taiwan is the upwind side to the CMR and receives most of the rainfall. As the western portion is at the leeward side, the rainfall amount was much smaller, where the lowest precipitation was found at central western Taiwan (deep blue area in Figure 10(a)). If TRaP was used, the estimated rainfall pattern (Figure 10(c)) is quite different with the ground rainfall truths, and the maximum accumulated rainfall also occurred in central western Taiwan (Figure 10(c)). However, a more actual rainfall distribution could be obtained if the I-TRaP was employed (Figure 10 (b)), exhibiting a high correlation coefficient (0.79) with the rain gauge measurements. Figure 11 shows another example regarding Typhoon Haitnag (2005). This typhoon crossed the northern CMR from eastern Taiwan. Due to the interaction between its counterclockwise rotation system and the CMR, the southern mountainous regions of Taiwan were at the windward side, which saw the maximum rainfall. The secondary rainfall areas were located in the upper portions of eastern Taiwan. At the opposite side, the downwind areas in northwestern and southeastern Taiwan received much smaller rainfall. These results once again reveal that the relative locations of the typhoon center with the CMR are one of the key factors affecting the rainfall distribution over the island. Again, the case of Typhoon Morakot (2009) also demonstrated a similar rainfall distribution to Typhoon Haitang (Figure 12). Typhoons Morakot and Haitang shared a very similar track when they struck Taiwan. The rain gauge rainfall for the two typhoons shows a very similar rainfall distribution too. However, the rainfall brought forth by Typhoon Morakot was much more catastrophic. It is worthy to note that the extreme heavy rainfall area over southern Taiwan's mountainous regions was predicted successfully via the I-TRaP technique (Figure 12). Although it still underestimated the actual precipitation amount, it exhibits a big improvement over the estimates by the original TRaP method. Meanwhile, another heavy rainfall area situated over center Taiwan was also detected by I-TRaP. For this event, the original TRaP completely failed in identifying such a region (Figure 12(c)) and the maximum accumulated rainfall location. By using the I-TRaP technique, the location is predicted to be situated on the southern mountainous Taiwan area. This is close to the station measured ground truth. However, the location is predicted to be located at the coastlines of southern Taiwan (black triangle) if the original TRaP was used. From these three examples, they all show the significance of the orographic effect on the typhoon rainfall, and the relative positions and orientation between the typhoon centers and the CMR play a very important role in the rainfall distribution over Taiwan. The result clearly demonstrates that the rainfall distribution could be predicted more accurately by considering the rainband rotation and the orographic effect when historical typhoon data sharing similar tracks were used (Figure 13).

(d) Sensitivity Test of Track Error

When establishing such rainfall potential techniques, the storm tracks are always assumed to be correct. However, this is considered an over-simplification. For actual applications, such as disaster mitigation plans, a complete preparation for different scenarios under varying possible tracks should be simulated. Therefore, a primary sensitivity test on the typhoon track error is conducted in this section. Typhoon Morakot is the selected tested case, and a 0.00 or 0.25 degree track error per day is added to the JTWC best track in longitude/latitude directions individually (Figure 14). Figure 14(e) is the original estimated result, while the results of the other figures are when track errors were added. It is apparent that the outcome does not change obviously, except for Figs.14(g) and (h) where the tracks deviate toward a southerly direction. In these two cases, the maximum rainfall values decreased significantly and the maximum rainfall location shifted southerly. Meanwhile, a similar change could be observed in Figure 14(d). Generally speaking, most of the figures show a very similar rainfall pattern and location as compared to Figure 14(e), meaning that the I-TRaP is not very sensitive to track errors. However, the track error simulated in this chapter is merely 0.25 degree per day. Larger track errors should be tested in future works.

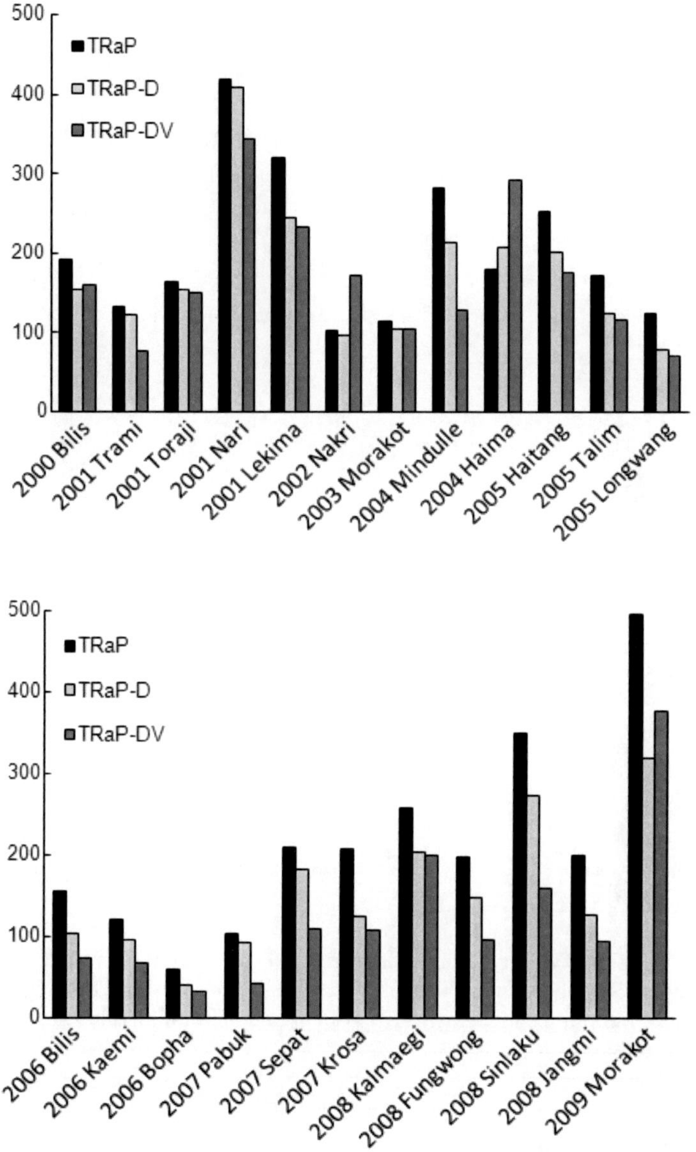

Figure 13. The RMSE change for 22 landfalling typhoons during 2000 to 2009. TRaP indicates the original TRaP result. TRaP-D indicates that the rainband rotation and redistribution step is added, and TRaP-DV shows the rescale step is added again (unit in mm).

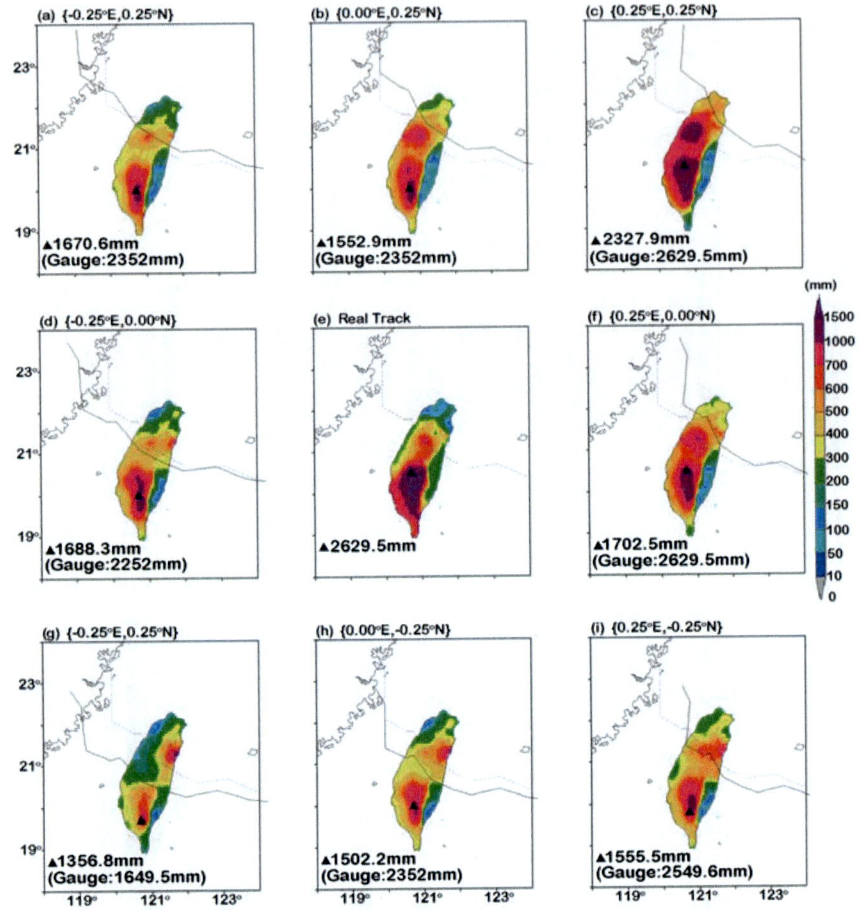

Figure 14. The sensitivity test of the track error. The dashed lines are the best tracks for Typhoon Morakot and the solid lines are tracks with track errors. The triangles indicate the maximum rainfall locations, where the accumulated rainfall value is shown in the lower left corner.

(e) Seasonal Winds

Located in the East Asia Monsoon region, the precipitation pattern of Taiwan is strongly affected by the southwest monsoon during the summer and by the northeast monsoon in the autumn and winter seasons. As the rainband circulation of typhoons often have a close interaction with such wind patterns, it can notably intensify the typhoon's rainfall activity. In the summertime, depending on the track of a typhoon's movement, the possible convergence between the typhoon's outer circulation and southwest winds could bring extreme rainfall to a particular area in or near southern or central Taiwan. For example, several such events have been witnessed over the southern Taiwan ocean region due to such convergence. More specifically, when a typhoon make landfall in northern Taiwan or passes offshore, the coupling effect will induce heavy rainfall over southern Taiwan (e.g. Morakot (2009))(Figure 15(a)). Meanwhile, if the typhoon track is further north of northern Taiwan, and assuming the southwest wind is also strong, the area of convergence will move upward to central Taiwan (e.g. Mindulle (2004)). On the other hand, in autumn if a typhoon moves in a more southerly direction and makes landfall in southern Taiwan or passes offshore, the typhoons' outer rainband circulations may interact with the northeast winds and bring intense precipitation over northern Taiwan (Figure 15(b)), especially the northeastern regions (e.g. Parma (2009) and Megi (2010)). Based on the typhoon cases that have struck Taiwan, many catastrophic typhoon rainfall events have been induced by such coupling effects. Hence, taking the seasonal winds into account is very crucial for establishing an accurate rainfall potential model for typhoons. By adopting a similar concept for the orographic effect step, the historical typhoon rainfall data under different seasonal winds could be categorized, and the effect from the seasonal winds on the precipitation amount can subsequently be parameterized. In this task, many parameters, such as the typhoon intensity, track and moving directions, seasonal wind speeds, water vapor content, and instability may need to be considered further.

5. Conclusion

Predicting the accumulated rainfall amount before a typhoon makes landfall poses a great challenge for weather forecasters and agencies, and is crucial for disaster mitigation. Although the original tropical rainfall potential technique is quite useful for forecasters to evaluate the rainfall potential, its assumptions are too simple to enable the technique to predict the rainfall pattern accurately. In addition, the technique cannot be applied to various complex terrain areas, such as Taiwan. Therefore, the I-TRaP method is proposed and demonstrated in this chapter. This new approach not only considers the rotation of the TC spiral rainband, it also takes the orographic effect into account via remapping and rescaling correction procedures. Instead of actually dealing with the terrain effect in a physical manner, the I-TRaP uses actual historical rainfall cases to delineate the orographic effect from Taiwan's CMR. The approximations are simple and easy to conduct, but the result reveals its effectiveness. The conclusions and suggestions drawn from this research can serve as a basis for relevant future works.

In this chapter, Ferraro's algorithm and SSM/I data were used to retrieve the rainfall. However, other rainfall algorithms and satellite data could be considered in future studies. An example is the NASA GPROF algorithm (Kummerow et al., 1996), which combines physical and statistical considerations, or other microwave data, such as the SSMI/S, TMI or AMSE-R data. Further, the usage of multiple satellite data can serve as an alternative choice. For example, data fusing of microwave and infrared images may provide rainfall retrievals with better spatial and temporal resolutions; incorporation of different satellite data with different weights could enable constructing an ensemble TRaP (e-TRaP) technique (Ebert et al., 2008).

The storm rotation speeds estimated from geostationary satellite images is helpful in improving the rainfall potential accuracy, which reveals that the storm rotation must be considered. Meanwhile, the rotation speeds may provide additional useful information to the storm intensity (Chao et al., 2010). Therefore, the maximum wind speed could also be estimated for disaster mitigation needs.

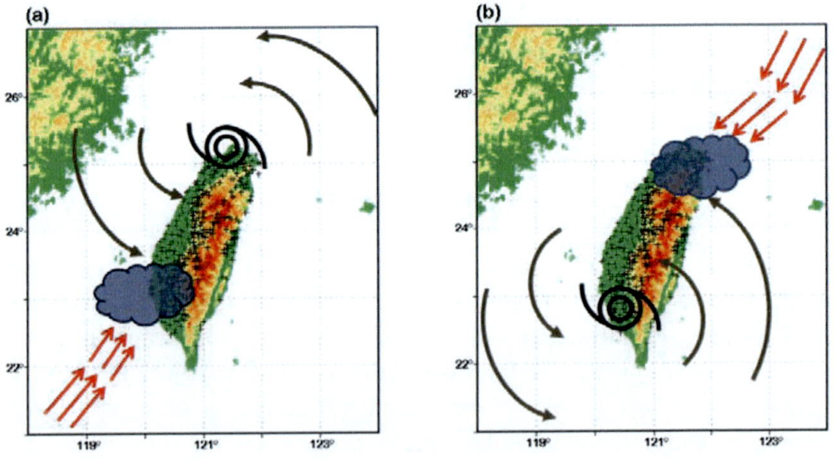

Figure 15. Interaction between the seasonal winds, typhoon outer circulation, and Taiwan's CMR. (a) for southwesterly wind cases, and (b) for northeasterly wind cases.

The accuracy of the predicted typhoon tracks is one key factor that affects the accuracy of the rainfall potential. Varying typhoon tracks from different forecast models or agencies can all be used to obtain a particular rainfall potential scenario under possible typhoon tracks.

A mesoscale topographic terrain could significantly affect a TC rainfall pattern. Fortunately, the rainfall pattern could be retrieved by the climatological TC rainfall pattern without performing too complex computations. The I-TRaP technique might also be easily conducted for other rugged terrain. Meanwhile, it is obvious that more historical cases would help to further categorize the typhoons into various intensities, moving directions, seasonal time frames and so forth; thereby attaining more accurate application results.

Coupling of the typhoon outer circulation and seasonal winds is another important influencing factor on the rainfall activity, which should be studied in future research. Technically speaking, the factor can be treated similarly as the orographic effect by using corresponding historical typhoon rainfall patterns under different seasonal winds to remap and rescale the rainfall rates. However, seasonal winds have different water vapor content and wind speeds. Hence, more typhoon cases need to be compiled in order to categorize them so as to determine the corresponding correction factors.

ACKNOWLEDGMENTS

Support of the National Science Council (NSC), Taiwan and great help of the Center Weather Bureau (CWB), Taiwan are gratefully acknowledged. We also wish to thank Mr. Guan-Ru Chen, Dr. Chun-Chieh Chao, Mr. Yu-Chun Chen, Mr. Cheng-Ju Chung, Miss Pei-Lun Hsieh, and Mr. Charlie, C. K., Liang.

REFERENCES

Alliss, R. J., S. Raman, and S.W. Chang, 1992: Special Sensor Microwave/Imager (SSM/I) observations of hurricane Hugo (1989). *Mon. Wea. Rev.*, 120, 2723-2737.

Brand, S., and J. W. Blelloch, 1974: Changes in the characteristics of typhoons crossing the island of Taiwan. *Mon. Wea. Rev.*, 102, 708-713.

Cheung, K. K. W., L.-R. Huang, and C.-S. Lee, 2008: Characteristics of rainfall during tropical cyclone periods in Taiwan. *Nat. Hazards Earth Syst. Sci.*, 8, 1463-1474.

Chao, C.-C., G.-R. Liu, and C.-C. Liu, 2011: Estimation of the upper-layer rotation and maximum wind speed of tropical cyclones via satellite imagery. *J. Appl. Meteor. Climatol.*, 50, 750-766.

Chiu, L.-S., G. R. North, D. A. Short, and A. McConnell, 1990: Rain estimation from satellites: effect of finite field of view. *J. Geophys. Res.*, 95, 2177-2185.

Dvorak, V. F., 1975: Tropical cyclone intensity analysis and forecasting from satellite imagery. *Mon. Wea. Rev.*, 103, 420-430.

Dvorak, V. F., 1984: Tropical cyclone intensity analysis using satellite data. NOAA Tech. Rep. NESDIS 11, pp.47.

Ebert, E., A. E. Salemi, M. Turk, M. Spampata, and S. Kusselson, 2009: Validation of the Ensemble Tropical Rainfall Potential (e-TRaP) for landfalling tropical cyclones. 16th Conference on Satellite Meteorology and Oceanography, AMS 89th Annual Meeting, Phoenix, AZ, USA, 11-15 January 2009.

Emanuel, K. A., 2005: Increasing destructiveness of tropical cyclones over the past 30 years, *Nature*, 436, 686-688.

Ferraro, R. R., 1997: SSM/I derived global rainfall estimates for climatological applications. *J. Geophys. Res.*, 102, 16715-16735.

Ferraro, R. R., P. Pellegrino, M. Turk, W. Chen, S. Qiu, R. Kuligowski, S. Kusselson, A. Irving, S. Kidder, and J. Knaff, 2005: The Tropical Rainfall Potential (TRaP) technique. Part II: Validation. *Wea. Forecasting*, 20, 465–475.

Gall, R., J. Tuttle, and P. Hildebrand, 1998: Small-scale spiral bands observed in hurricanes Andrew, Hugo, and Erin. *Mon. Wea. Rev.*, 126, 1749–1766.

Grody, N. C., 1991: Classification of snow cover and precipitation using the Special Sensor Microwave/Imager (SSM/I), *J. Geophys. Res.*, 96, 7423-7435.

Hollinger, J., R. Lo, G. Poe, R. Savage, and J. Pierce, 1987: Special Sensor Microwave/Imager user's guide. Naval Research Laboratory, Washington, D.C., pp.120.

Kidder, S. Q., M. D. Goldberg, R. M. Zehr, M. DeMaria, J. F. W. Purdom, C. S. Velden, N. C. Grody, and S. J. Kusselson, 2000: Satellite analysis of tropical cyclones using the Advanced Microwave Sounding Unit (AMSU). *Bull. Amer. Meteor*. Soc., 81, 1241–1259.

Kidder, S. J. Kusselson, J. A. Knaff, and R. J. Kuligowski, 2001: Improvements to the experimental tropical rainfall potential (TRaP) technique. Preprints, 11th Conf. on Satellite Meteorology and Oceanography, Madison, WI, Amer. Meteor. Soc., 375–378.

Kidder, S. Q., S. J. Kusselson, J. A. Knaff, R. R. Ferraro, R. J. Kuligowski, and M. Turk, 2005: The tropical rainfall potential (TRaP) technique. Part I: Description and examples. *Wea. Forecasting*, 20, 456-464.

Kummerow, C., W.S. Olson, and L. Giglio, 1996: A simplified scheme for obtaining precipitation and vertical hydrometeor profiles from passive microwave sensors. *IEEE Trans. Geosci. Remote Sens.*, 34, 1213-1232.

Lin, Y.-L., S. Chiao, T.-A. Wang, M. L. Kaplan, and R. P. Weglarz, 2001: Some common ingredients for heavy orographic rainfall. *Wea. Forecast*ing, 16, 633-660.

Liu, C.-C., 2009: The influence of terrain on the tropical rainfall potential technique in Taiwan. *Wea. Forecasting,* 24, 785-799.

Liu, G.-R., C.-C. Chao, and C.-Y. Ho, 2008: Applying satellite-estimated storm rotation speed to improve typhoon rainfall potential technique. *Wea. Forecasting*, 23, 259-269.

Lonfat, M., F. D. Marks Jr., and S. S. Chen, 2004: Precipitation distribution in tropical cyclone using the Tropical Rainfall Measuring Mission (TRMM) microwave imager: A global perspective. *Mon. Wea. Rev.*, 132, 1645-1660.

Pfost, R. L., 2000: Operational tropical cyclone quantitative precipitation forecasting. *Natl. Wea. Dig.*, 24, 61–66.

Ramsay, H. A., and L. M. Leslie, 2008: The effects of complex terrain on severe landfalling tropical cyclone Larry (2006) over Northeast Australia. *Mon. Wea. Rev.*, 136, 4334-4354.

Riehl, H., and J.S. Malkus, 1961: Some aspects of hurricane Daisy, 1958. *Tellus*, 13, 181–213.

Senn H. V., and H. W. Hiser, 1959: On the origin of hurricane spiral rain bands. *J. Meteor.*, 16, 419–426.

Wang, S.-D., C. L. Yen, T.-J. Chen, and S. L. Shie, 1986: The rainfall property and forecast of typhoons in Taiwan (in Chinese). Technical Report of Disaster Mitigation Research (3), National Science Council, No.74-51, pp.152, Taiwan.

Weng, F., and N. C. Grody, 1994: Retrieval of cloud liquid water using the Special Sensor Microwave Imager (SSM/I). *J. Geophys. Res.*, 99, 25535-25551.

Wimmers, A. J., and C. S. Velden, 2007: MIMIC: A New Approach to Visualizing Satellite Microwave Imagery of Tropical Cyclones. *Bull. Amer. Meteor. Soc.*, 88, 1187–1196.

Wu, C.–C., T.–H. Yen, Y.–H. Kuo, and W. Wang, 2002: Rainfall simulation associated with typhoon Herb (1996) near Taiwan. Part I: The topographic effect. *Wea. Forecasting*, 17, 1001-1015.

Yeh, T.-C., and R. L. Elsberry, 1993: Interaction of typhoons with the Taiwan orography. Part I: Upstream track deflections. *Mon. Wea. Rev.*, 121, 3193-3212.

Zhang, F., Y. Weng, Y. -H. Kuo, J. S Whitaker, and Baoguo Xie, 2010: Predicting typhoon Morakot's catastrophic rainfall with a convection-permitting mesoscale ensemble system. *Wea. Forecasting*, 25, 1816-1825.

In: Rainfall Forecasting
Editor: T.S.W. Wong, pp. 233-273
ISBN: 978-1-61942-134-9

Chapter 6

REAL TIME FLOOD CONTROL USING QUANTITATIVE PRECIPITATION FORECAST AND ENSEMBLE APPROACH

Oliver C. Saavedra V. [*,1,2], ***Toshio Koike*** [3,†], ***Masahiro Ryo*** [1] ***and Shinjiro Kanae*** [1]

[1] Tokyo Institute of Technology, 2-12-1, Ookayama. Meguro-ku, Tokyo, Japan.

[2] Egypt-Japan University of Science and Technology, New Borg El-Arab, Egypt.

[3] University of Tokyo, 7-3-1, Hongo, Bunkyo-ku, Tokyo, Japan.

ABSTRACT

This chapter discusses the feasibility of using quantitative precipitation forecasts (QPF) at an operational stage for structural and nonstructural flood control. It introduces the ensemble approach to assess forecast inaccuracy, providing valuable decision support information. Comparison of recent forecasts with real-time, in situ measurements can be used to determine forecast bias. Since underestimated or overestimated

[*] Emails: saavedra.o.aa@m.titech.ac.jp, ryo.m.aa@titech.ac.jp, kanae@mei.titech.ac.jp
[†] Email: tkoike@hydra.t.u-tokyo.ac.jp

QPF might occur in areas surrounding the basin, alternatives to define the error are introduced. Then, spatial analysis is performed, not only within the targeted basin, but also within nearby areas. To assess the intrinsic uncertainty of forecasts, an ensemble method based on noise perturbation is proposed. To attempt an appropriate ensemble spread, a weighting system is developed according to an evaluation of spatial and temporal bias. The generated precipitation members are used to force distributed hydrologic models, and to obtain forecasts of reservoir inflow for structural flood control. Similarly, in nonstructural flood control mode, the precipitation ensemble can be generated for flood forecasts at the control point. Then, two types of systems can be distinguished, for supporting dam operation and early flood warning. These systems can evaluate precipitation forecast performance from previous time steps, to support proper and timely decisions. This chapter also treats study cases of operational river management in the Upper Tone Reservoir system in Japan, in which mesoscale forecast uncertainty was inserted into the dam release schedule. Additionally, the development of an early flood warning system is introduced, using global and regional forecasts in the Huong River basin of Vietnam. The implications of both cases are discussed, and recommendations for more generalized cases are highlighted.

Keywords: qpf, rainfall forecast, hydrologic modeling, ensemble members, flood control, Tone River, Huong River, real-time dam operation, flood warning, inflow forecast, stream flow forecast, error forecast.

1. Introduction

In this section, the ongoing threat of floods to society will be discussed, as well as improvements in prediction and forecast techniques. These improvements require better understanding of the phenomena and real-time measurements. Examples for these cases are summarized.

1.1. Floods as a Threat to Society, Current and Future Expectations

Floods are natural events that have always been an integral part of the geologic history of Earth. According to Berz (2000), flood disasters account for a third of all natural disasters by frequency and economic loss, and they cause more than half of all fatalities globally (see Figure 1). Floods have caused the greatest number of deaths in the poor and heavily populated countries of the world, particularly in Asia. An example of recent severe flooding was the 2010 Indus River Basin flood in Pakistan (see Table 1). The government reported that the number of afflicted people was approximately 20 million, and that the death toll was close to 2,000. Flood hazards may intensify by 2050 because of global warming, according to the fourth assessment report of the Intergovernmental Panel on Climate Change (Team et al., 2007). Heavy rainfall that causes floods can be attributed to large water cycle variations at global and regional scales, whereas flood damage occurs at the basin scale. The frequency of these events might increase mainly over humid Asian regions, and over tropical areas of Africa and South America, consistent with increases of atmospheric water vapor. However, some areas of the world are projected to have decreased heavy rainfall in the future, such as most parts of North America and northeastern Europe (Hirabayashi et al., 2008)

1.2. Current Quantitative Precipitation Forecasts (QPF) at Operational Stage

Among the most significant scientific advances of the past century is our ability to simulate complex physical systems with numerical modeling, thereby predicting their evolution. An outstanding occurrence has been the development of general circulation models of the atmosphere and ocean, which have brought two great advantages: (1) it is now possible to predict the weather several days in advance, with a high degree of confidence; and (2) we are gaining better understanding of the factors changing our climate, and their likely timing and severity. It is no exaggeration to describe the advances of the past half century as revolutionary. Thanks to these advances, meteorology is now firmly established as a quantitative science. Its value and validity are demonstrated daily by the acid test of any science, its ability to predict the future (Lynch, 2008). Operational quantitative forecasting today uses guidance from a wide range of models. In most centers, a blend of global and local models is used. Weather forecasting centers using numerical weather prediction (NWP) are listed in Table 2.

The accuracy of weather forecasting at the basin scale has especially improved in recent years, owing to more reliable weather prediction models. Precipitation is one of the most difficult weather variables to predict, because the atmosphere can be very unstable. However, advanced techniques have enabled reasonable predictions at regional scale (e.g., Golding, 2000;

Krzysztofowicz et al., 2004; Honda et al., 2005). Since precipitation is the main input data for hydrologic models, QPF accuracy is reflected in the streamflow forecast. Recent works (Jasper, 2002; Collischonn et al., 2005; Saavedra et al., 2009) have shown the feasibility of forcing distributed hydrologic models (DHMs) with mesoscale QPF, to simulate hydrologic processes at the basin scale.

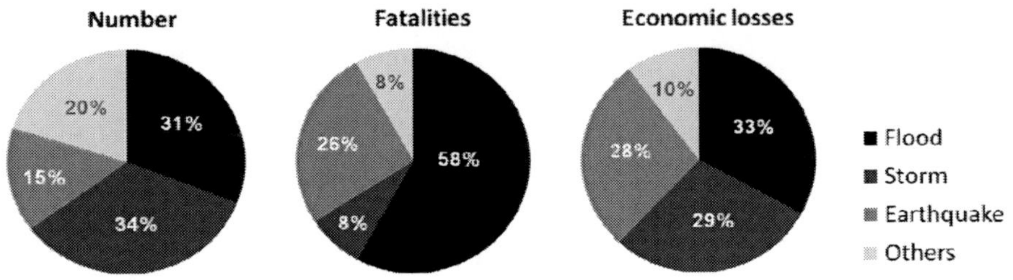

Figure 1. National catastrophes worldwide (1988-1997).

Table 1. Recent catastrophic flood events causing the largest numbers of cost losses all over the world from 1980 to 2010 (NetCatSERVICE, 2011)

Period	Country / region	Overall losses	Fatalities
May-Sep. 1998	China	30,700	4,159
Jun-Aug. 1996	China	24,000	3,048
Jun-Aug. 1993	USA	21,000	48
Aug. 2002	Europe	16,500	39
Jul-Aug. 1995	North Korea	15,000	68
May-Sep. 1991	China	13,600	2,628
Jun-Sep.1993	China	11,000	3,300
Jun. 2008	USA	10,000	24
Jul-Aug. 2010	Pakistan	9,500	1,760
Nov. 1994	Italy	9,300	68
		(US$ m, original values)	

2. Flood Control Tools

The two main options for reducing flood damage impacts on society are introduced. These are so-called structural and nonstructural measures.

Table 2. Summary of centers providing Quantitative Precipitation Forecast

Weather Forecasting Center	NWP started year
The European Centre for Medium-Range Weather Forecasts (ECMWF)	1975
National Centers for Environmental Prediction (NCEP)	1954
National Oceanic and Atmospheric Administration (NOAA)	1970
Japan Meteorological Agency (JMA)	1959
UK Met Office (UKMO)	1977

2.1. Structural and Nonstructural Flood Control Systems

Flood damages can be reduced by appropriate control systems, which can be divided into structural and nonstructural measures. Examples of the former group are the construction of flood-control facilities that account for flood characteristics; these facilities are represented by reservoirs, diversions, levees or dikes, and channel modifications (Mays, 2005). The nonstructural measures do not entail any construction, and modify the damage susceptibility of floodplains. They are designed to avert flood damage for permanent facilities, and provide for reduced potential damage during flood events. Nonstructural measures include land use controls, inundation maps, and flood warning. Proper flood management may require a combination of engineering structures, nonstructural and administrative measures, for minimizing flood losses at the basin scale. Structural or nonstructural countermeasures are mainly applied to implement appropriate flood control. Both are illustrated in Figure 2.

2.1.1. Structural

The purpose of flood control is to reduce or alleviate the negative consequences of flooding, using structural measures such as reservoirs, diversions, levees, and channel modifications.

7. Reservoirs are used to store river discharge, causing flooding and release after the flood event, thereby diminishing the magnitude of peak discharge. Reservoirs lower the flood-flow frequency curve by decreasing peak discharge in a specific event.
8. Diversions are used to reroute river discharge, for lessening peak flows during flood events. Diversions are designed to modify the

frequency curve, so that flow magnitude for a specific event is decreased.

9. Levees or dikes are used to keep flood flows from floodplain areas, where damage can occur. Levees reduce damage in protected areas from water surface stages within the main channel. Constricting the flow within levees reduces the natural storage in flood waves.

The effects of reservoir, diversion and levee over stage, discharge cost and exceedance frequency can provide effective structural flood control, as seen in Figure 3. Flood damage reduction measures can be expressed by basic assessment functional relationships, as in the figure. The stage-damage relationship defines flood severity in terms of damage cost at various stages. The stage-discharge relationship is better known as rating curves. Flood–flow frequency relationships indicate the recurrence of events by flood magnitude (Mays, 2005).

The degree of flood peak reduction depends on reservoir storage capacity. The water level exceeding cross-sections can estimate inundation areas and flood damage. The exceedance probability vs damage is seen in Figure 4.

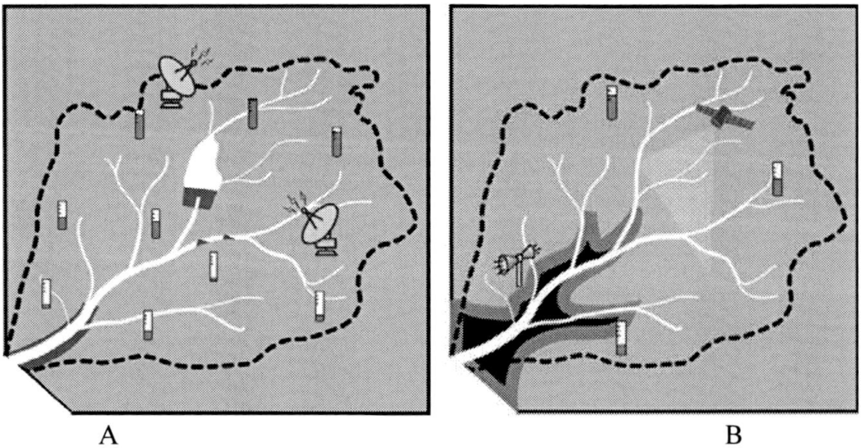

Figure 2. Structural (A) and nonstructural (B) flood control systems. At former case a reservoir might efficiently reduce the inundated areas downstream using radar while in latter a timely flood warning is emitted for evacuation from flooded areas using remote sensing techniques besides rain gauges.

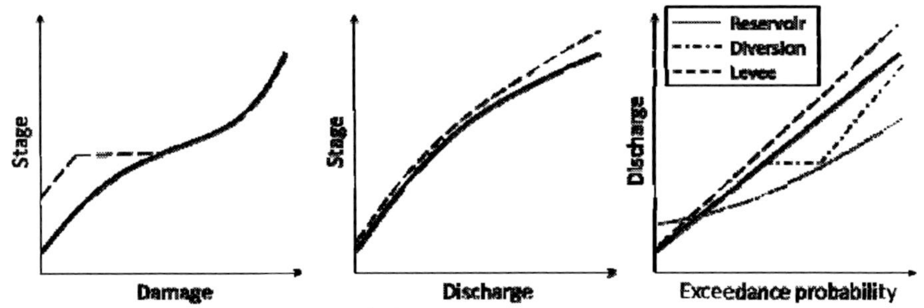

Figure 3. Effect of flood control alternatives by different structural measures. The solid lines depict flood assessment functional relationships. The left graph shows that only the levee might modify the stage-damage relationship. The center graph also illustrates the potential effect on the levee rating curve. The right graph shows that flood-flow frequency can modify the discharge using three alternative structures as the three discontinuous lines shown. Graphs adapted from Mays (2005).

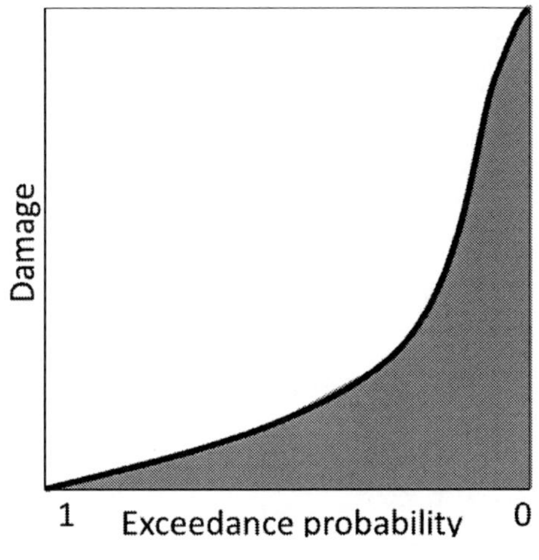

Figure 4. Exceedance probability versus damage.

The Operational dams for flood control and other purposes require pre-established rule curves for normal events (Chang and Chang, 2001). However, during extreme flood events, dam operators must exercise their own judgment. The coupling of optimization algorithms with hydrologic models to improve release schedules has been reviewed (e.g., Yeh, 1985; Labadie, 2004). Ongoing research has shown improvements when using operational forecasts and real-time data measurements.

2.1.2. Nonstructural

Nonstructural measures reduce potential damage without rigid structures, as described previously. However, they might include improvements to permanent structures, enhancing durability and efficiency. These measures can be divided mainly into:

- Flood proofing
- Flood warning (hazard mappping, warning system)
- Land-use

Flood proofing is designed to modify the damage potential of individual structures susceptible to flood damage by means of

- Elevating structures
- Waterproofing exterior walls
- Rearrangement of structural working space

Flood proofing is most desirable for new facilities, and alters only the stage-damage cost relationship. This is illustrated in Figure 5, which defines flood severity in terms of monetary cost. Performance can be ensured, for example, by adequate installation of closures and confirming the stability of levees.

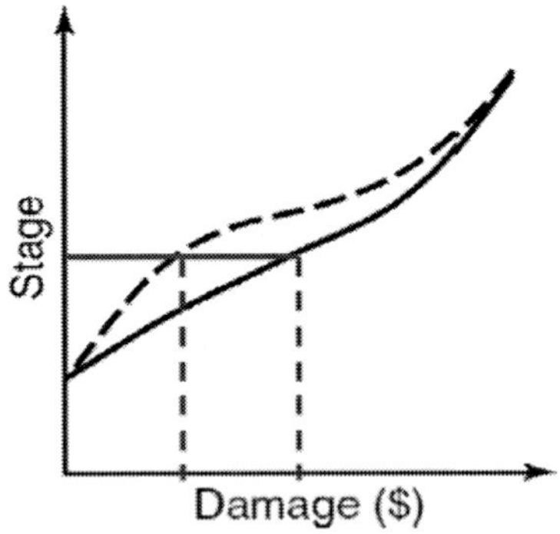

Figure 5. Stage-damage relationship as a nonstructural measure example. The solid curve depicts the flood severity in terms of damage monetary cost. The dashed curve depicts the flood proofing effect. It can be seen that at the same stage flood proofing can successfully reduce the damage cost following horizontal line. Adapted from Mays (2005).

A Flood Warning System (FWS) reduces the flood damage by giving public an opportunity to act before flood stages reach damaging levels. Such a system may results in:

- Reduced inundation damage
- Reduced cleanup costs
- Reduced cost of service disruption
- Reduced costs because of decreased health hazards

The crucial points of FWS plans are

- An FWS is a critical component of other flood-damage-reduction measures
- An FWS provides notice with lead-time to floodplain occupants, for reducing potential damage
- Lead-time provides an opportunity to elevate contents of structures, to perform minor proofing, and to remove property susceptible to flooding
- The greatest value of flood warnings is reducing or eliminating loss of life
- Flood warning requires real-time flood forecasting and communication facilities to warn floodplain inhabitants

Land use controls deal with administrative actions for modifying floodplain land use, to make it compatible with the potential flood hazard:

- Zoning and other building ordinances
- Direct acquisition of land and property
- Building codes
- Flood insurance
- Information programs (local, prefecture, national government)

2.2. Precipitation Forecast Error and Bias

To account for QPF errors in terms of intensity and displacement of the main rainfall pattern, Ebert and McBride (2000) proposed contiguous rain areas. It would be helpful to locate these areas with respect to the targeted basin. Underestimated or overestimated forecast precipitation can occur in areas surrounding the basin. Analysis should be performed in different zones. To estimate the forecast error between QPF and real-time observed precipitation, that error may be expressed as (1) absolute error, (2) relative error, (3) root mean square error, and (4) ratios. The following gives a detailed explanation of these error measurements.

(1) Absolute error

This error is the most basic method for assessing forecast bias. Observed precipitation may be treated as the true value. Then, the definition of absolute error is the subtraction of observed from forecast values, as

$$\Delta\_error \equiv Forecast - Observed \qquad (1)$$

(2) Relative error

This error is generally used to highlight the tendency (bias) of the forecast, and is defined as absolute error over observed. This indicates the comparative accuracy of the forecast, and the expression can be useful for comparison of error at different magnitudes. However, it cannot be applied when the observed value is zero.

$$\delta\_error \equiv \frac{\Delta\_error}{Observed} = \frac{Forecast - Observed}{Observed} = \frac{Forecast}{Observed} - 1 \qquad (2)$$

(3) Root mean square error

Root mean square error (RMSE) represents the degree of forecast error. It allows calculation of the forecast accuracy itself. There is an advantage in that RMSE can be calculated from various measurements, and it is defined as

$$RMSE \equiv \sqrt{\frac{\sum (Forecast - Observed)^2}{N}} \qquad (3)$$

(4) Ratio

In cases where the forecast error bias fluctuates considerably, the error can be defined as a simple ratio of forecast to observed precipitation, as

$$Ratio \equiv \frac{Forecast}{Observed} \tag{4}$$

Once the appropriate error expression is defined according to the magnitudes and number of samples, a real-time evaluation of QPF is desirable. This evaluation compares recently issued forecasts against real-time observations, as shown in Figure 6. This comparison is valid until the bold vertical line represents the present time. After that, it is only possible to have the QPF forecast according to the length of the lead-time. If the total lead-time is divided by the issuance frequency, interval frames such as t0 and t1 are obtained. These intervals might be used to define the periods of the QPF evaluation with onsite measurements. According to requirements and available data, t0, t−1, t−2, t−3, and so on could be used for evaluating recent QPF issuances.

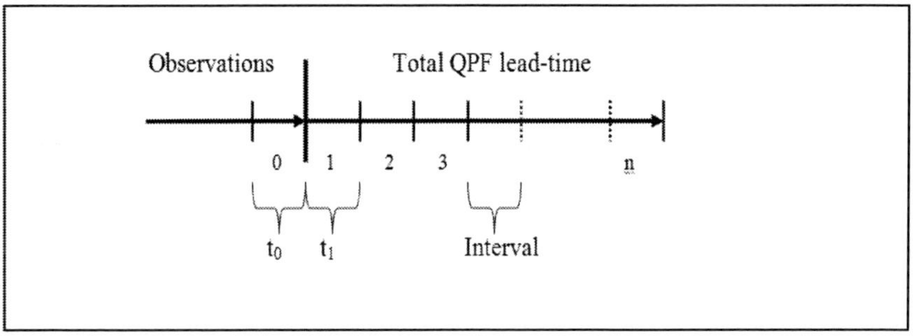

Figure 6. Evaluation forecast error. In real time basis only recent observations and short term QPF, normally hours, is available. The total QPF lead time can be divided by ***n*** intervals according to the issuance frequency so-called t1, t2, and so on. The evaluation of forecast error can be done regressively from t0, t-1, t-2 and so on. The bold vertical line represents the present time.

The spatial variability of the error QPF bias compared with observations may be located outside the basin area. This means overestimation or underestimation magnitudes of QPF, which represent the event, might lie outside the watershed. Therefore, it is crucial to identify proximity to the water divide. The use of circular buffers is suggested, as shown in Figure 7. These can easily be created using a geographic information system (GIS). These buffers are drawn using the gravity mass of the watershed as the center point. The first buffer covers the entire watershed. The second, third and successive ones were defined using an incremental ratio between 10 and 20%. Additionally, the window encompassing all buffers can be defined as a rectangular shape, as shown in Figure 7. In this fashion, spatial analysis is performed not only within the targeted basin, but also within nearby areas.

To reflect the intrinsic uncertainty of forecasts, an ensemble method based on noise perturbation is proposed. To attempt an appropriate ensemble spread, a weighting system according to the spatial and temporal bias evaluation is required. The outputs are perturbation weights for use over the total lead-time forecast, to generate the ensemble precipitation members.

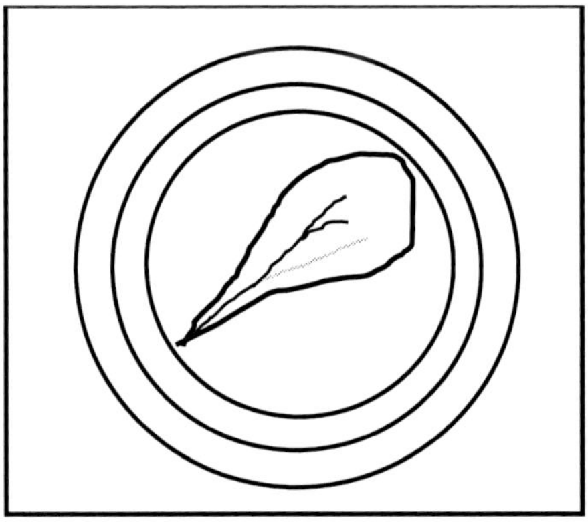

Figure 7. Evaluating forecast error using buffers around the target watershed. This is critical important to track missplaced events within nearby areas the basins, and it could eventually being considered for decision making at basin scale river management.

2.3. Ensemble Method to Express QPF Uncertainty

Because of the increase in computation ability over the past 15 years or so, ensemble forecast techniques known as ensemble prediction systems (EPS) have been used around the world. These EPS account for meteorological uncertainties. EPS allow for the prediction of the most likely scenario, and for the probability of occurrence of other events. Forecast uncertainty arises from each process within a NWP system, such as observation and data collection, data assimilation and modeling. These uncertainties can be treated by EPS. One of the main purposes of numerical ensemble forecasts is to assimilate initial uncertainty (initial error) and forecast uncertainty (forecast error), using either the initial perturbation method or the multi-model/multi-physics method. In fact, the mean of an ensemble forecast generates a better forecast than a deterministic forecast, after a short lead-time in global modeling applications. Furthermore, ensemble forecasts can deliver a probabilistic forecast to users based on the probability density function, instead of a single-value forecast from the traditional deterministic system. There are presently two major methods to generate an ensemble model forecast. One is the initial perturbation method, which adds small perturbations to the initial analysis. Another set of ensemble forecasts is produced using different numerical models (spectrum and grid models) and different physical packages. These aspects make an EPS forecast attractive for flood forecasting systems, and afford the potential to extend lead-time and better quantify predictability. Many flood forecast systems rely on precipitation inputs, which come initially from observation networks composed of rain gauges and radar. However, to extend forecast lead-time, NWP models combined with EPS can be used to assimilate available onsite measurements. The generated precipitation members are then input to DHMs (Figure 8) to forecast reservoir inflow streamflow at flood control points.

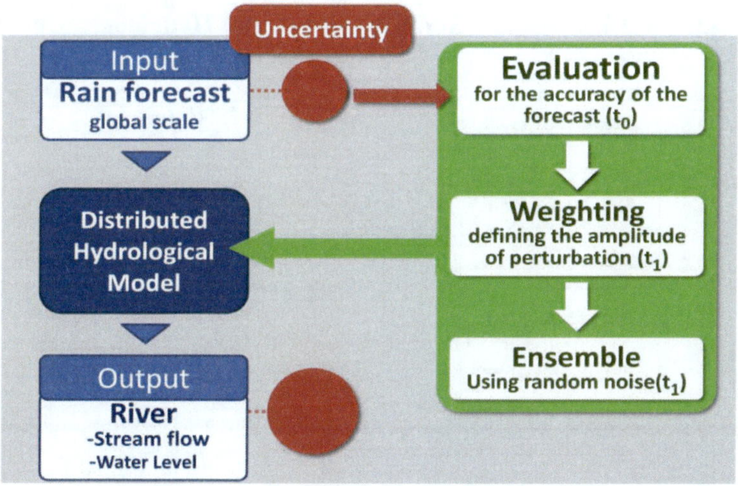

Figure 8. General scheme to assess uncertainty of QPF by using available ground observations. The spread of the ensemble precipitation is defined by a weighting approach from comparison at previous time steps. Then, that information is reflected to generate different members of stream flow using distributed hydrologic models.

2.4. Hydrologic Model

A mathematical or numerical hydrologic model is discussed in this section. An introduction to hydrologic models is first, followed by their classification. Subsequently, the geomorphology-based hydrologic model (GBHM) and water-energy-budget (WEB)-DHM models are described.

A mathematical hydrologic model should be understood as a simplified representation of the real water cycle. Its main objective is to simulate hydrologic processes mathematically and numerically as close to reality as possible, enabling accurate predictions for areas with no measurements, or for new events. The main applications are flow forecasting, water yield, point/non-point pollution sources, and flood control. Such a model can be a tool for water resource management, or for the design and operational stages of hydraulic projects, such as dams, urban drainage, and so on.

According to mathematical implementation, two main hydrologic models can be distinguished: stochastic and deterministic. Both types of models are shown in the first column of Table 3.

Table 3. Classification and Development of Hydrologic Models

	Type	Year	Modeler Name (s)	Improvements	Model Name
STOCHASTIC MODELS		1970	Box and Jenkins	Time series record Probability distribution variables	Stochastic
Derministic models	Black box	1850	Mulvaney	Peak Flow Discharge: Qp = CiA	Rational Method
		1932	Sherman	Superposition principle, volume of surface runoff	Unit Hydrograph
	Lumped	1966	Crawford and Linsley	Incorporate simple physics at averaged catchment	Stanford Model IV
		1967	Sugawara	Simple and easy to understand, and flexible to add modules	Tank Model
		1983	Leavesley, Lichty, Troutman, and Saindon (USGS)	Distributed conceptual model	PRMS
	White box	1969	Freeze and Harlan	Basic understanding of physical processes as non-linear	Freeze and Harlan Blueprint
		1979	Beven and Kirkby	Use of Finite Element methods Flow routing by topography.	TOPMODEL
		1986	Abbott (Systéme Hydrologique Européen)	Fully modular system for mathematical description	SHE

		Year	Modeler Name (s)	Improvements	Model Name
		1989	Voeroesmarty	Macro-scale models	Voeroesmarty Model
		1992	E. Wood	Continental-scale model	VIC model
		2001	DHI water and environment	Physically-based using computing grids enhanced groundwater, and other modules	MIKE-SHE
		2002	D. Yang et al.	Physically-based using flow intervals; then fast computing	GBHM model
		2009	L. Wang et al.	Water and energy budget model using flow intervals	WEB-DHM model

2.4.1. Stochastic Models

These models are derived from time series analysis of historical records. They use variables dealing with a probability distribution, which permits future predictions. Therefore, they can also be understood as empirical models.

2.4.2. Deterministic Models

Unlike stochastic models, the variables in these models are free from random variation, and have no distribution of probability. Therefore, an input can always produce the same output. Deterministic models can be mainly classified into three categories: empirical (black box), conceptual (lumped), and physically-based, distributed (white box) models.

2.4.3. Empirical (Black Box) Models

These models are based on transfer functions by mathematical equations that relate inputs with outputs. These models do not have any physical basis. A good example in this category is the unit hydrograph.

2.4.4. Conceptual (Lumped) Models

These models occupy an intermediate position between an empirical black box analysis and a full, physically based approach. They are formulated on a semi-empirical basis in which the catchment is averaged; therefore, no spatial distribution is considered. For example, the Precipitation-Runoff Modeling System uses homogeneous hydrologic response within a hydrologic response unit, and the watershed is divided into these units (Carey and Simon, 1984).

2.4.5. Physically-based Distributed Hydrologic (White Box) Models

Physically-based distributed models consider spatial variations of all variables and parameters in the overall catchment. Then, they describe the catchment using the physics of water flow processes, relying on numerical methods of the governing equations: mass, energy and momentum conservation. As can be imagined, these models more closely approximate the real hydrologic processes in a catchment, compared with the models described above. Examples in this category are the TOPMODEL (Beven and Kirkby, 1979), SHE (Abbott, 1986), or MIKE SHE models (Madsen and Jacobsen, 2001).

White box models also have limitations, however. For example, they have great computational expense, because of the large amount of data required and the need for parameter values at the element scale. To overcome this input requirement, assumptions are usually made.

The selection of a model from those described above will depend on the objective of the modeling, data and time availability, budget, computer resources, and familiarity with the software. In general terms, white box models can be applied to any hydrologic processes, but at relatively high expense in time and money. Therefore, traditional simple models are still in use, because of their simplicity and inexpensiveness.

During recent decades, physically-based DHM have seen the most development of all hydrologic models. This is because of a growing demand for forecasting spatially variable effects, increased computing power, and user-friendly software development. Moreover, visualization of results has improved because of the availability of remote sensing data and the utility of GIS for spatial data analysis.

Remote sensing data, such as satellite imagery, currently furnishes spatial information that enhances features of the hydrologic cycle, like topography, land use, and vegetation. Furthermore, distributed rainfall amounts are estimated by active radar systems.

The application of GIS in hydrologic modeling has proven to be a powerful tool in spatial data preparation from different sources and scales. Moreover, hydrologic catchment properties can be obtained from satellite image processed data, and evaluated with GIS. Examples of such properties are topography (DEM), river networks, soil maps, land use/cover maps, and so on (Figure 9).

In recent years, continental and even global scale hydrologic modeling has begun. Moreover, coupled hydrologic models have been developed. For example, coupled atmospheric and land surface models (LSMs) have promoted integral understanding. In addition, virtual representation of hydrologic processes is possible, running in 'real-time' with data assimilation.

Geomorphology-Based Hydrologic Model (GBHM)

The GBHM proposed by Yang (2002) can be used to simulate spatially distributed hydrologic processes. It is a grid-based model, in which the computational unit is a geometrically symmetric hillslope. This element is viewed as a rectangular inclined plane, with a defined length and unit width. The inclination angle is given by the surface slope, and the bedrock is assumed parallel to the surface. Because the GBHM is a physically based model, it solves the governing equations: continuity, momentum, and energy equations, using two modules.

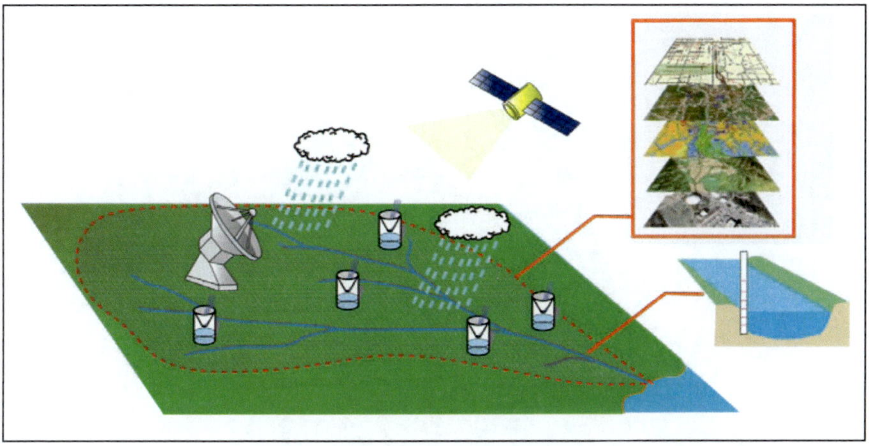

Figure 9. Application of remote sensing (e.g. satellite-based, ground-based radar) with on-site measurements to simulate hydrological processes using distributed hydrological models. Spatial data are shown in GIS layers.

First, a hillslope module evaluates hydrologic processes such as canopy interception, evapotranspiration, infiltration, and surface flow, as well as exchanges between groundwater and surface water. The vertical plane is divided into several layers, including canopy, soil surface, unsaturated zone, and groundwater. Actual interception is determined by precipitation amount and deficit of canopy water storage. Actual evapotranspiration is calculated as evaporation from canopy water storage, transpiration from the root zone, evaporation from surface storage, and evaporation from the soil surface. The unsaturated zone is divided into layers, and water flow is governed by Richard's equation. The equations for the saturated zone are mass balance and Darcy's law. Second, water routing in the river network is determined using one-dimensional kinematic wave equations, which are solved by finite-difference approximation. This simulation sequence is performed at each sub-basin. Moreover, each sub-basin is divided into numerous flow intervals by considering the flow distance to the outlet. The flow interval length is defined as twice the simulation grid-cell size. The hillslope module updates the surface runoff and subsurface flow, which become the lateral inflow into the main river channel. The river flow routing module then conducts water gathered from each flow interval from the source to the outlet, along the direction of the main river channel. This is lumped within each sub-basin.

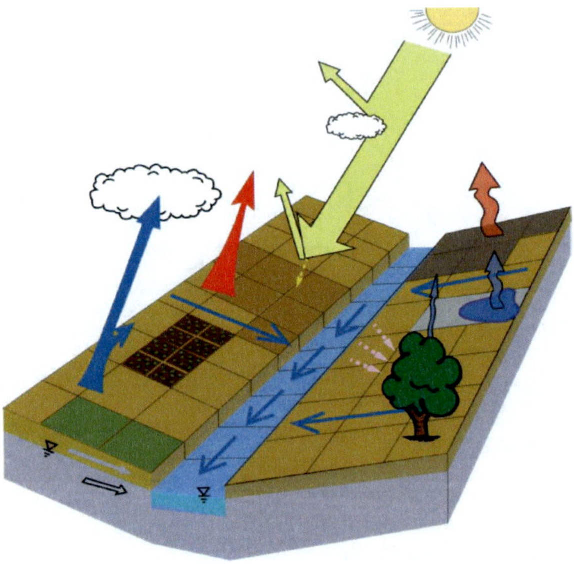

Figure 10. Scheme of water and energy budget approach at hillslope unit used by WEB-DHM model. At each unit, it can be assigned different land use classes such paddy rice, forest, bare soil. The partitioning of solar radiation is indicated in yellow arrows while evaporation from soil in red bended arrow, and from water bodies in blue bended one. Transpiration from canopy and root zone are seen as thin blue arrow above tree. The latent heat of transpiration seen in double blue arrow below cloud, and sensible heat in red straight one.

Distributed Biosphere Hydrologic Model (DBiHM)

The GBHM, like most DHMs, can assimilate rainfall patterns and estimate fluxes such as a stream flow, at selected points. They also provide spatial distributions of soil moisture content, evaporation, and transpiration. However, detailed energy exchange between the land surface and lower atmosphere is neglected in most DHMs (e.g., Yang et al., 2001). Thus, combining hydrologic and LSMs improves the representation of water and energy cycles within a basin. The simple biosphere model SiB2 (Sellers et al., 1996), a well-known LSM, has been embedded in GBHM to produce a water- and energy-budget-based DHM, namely WEB-DHM (Wang et al., 2009). This model was validated in humid areas and can predict the evolution of land surface temperature and soil moisture. The SiB2 submodel calculates turbulent fluxes between the lower atmosphere and land surface independently, as shown in Figure 10. It then updates the surface runoff and subsurface flow, which become the lateral inflow to the main river channel. Next, the GBHM submodel simulates water flow within the river network, using one-dimensional kinematic wave equations solved by finite-difference approximation. This simulation sequence is performed for each sub-basin and then at flow intervals, based on the flow distance to the outlet. The flow intervals are linked by the river network, based on geomorphology. The flow downstream may be affected by the operation of upstream dams. A dam function based on the mass balance approach is used to express the change in volume. Inflows to dams are simulated by the GBHM submodel.

The GBHM or WEB-DHM can express spatial variability of the hydrologic processes within a basin, according to its weather characteristics and setting. For example, for those basins where potential evaporation is greater than precipitation, WEB-DHM can simulate heat fluxes. Overall, WEB-DHM uses the governing equations more comprehensively, which may better simulate soil moisture. This is critically important for initial conditions, before flood evolution. However, a greater amount of input data, more expertise and calibration are required.

2.5. Optimization Algorithm for Dam Operation

Effective flood control can be achieved by improving dam operation. This can be done by coupling an optimization algorithm. Then, proper release rates can be forecast beforehand, to be applied during and after a flood event. The heuristic algorithm shuffled complex evolution (SCE; Duan et al., 1992), which searches for global optima, was chosen. The efficiency and robustness of SCE has been demonstrated by comparison against a variety of methods (e.g., Cheng et al., 2005; Saavedra, 2010a). The algorithm has been widely used in the calibration of hydrologic models, but here the application is for dam operation. First, a sample of points is distributed stochastically over the feasible decision variable space, by defining the lower and upper bounds of each variable to be optimized. Every point is meant to represent a member of a population, for which only the best fit will be kept. Each individual is characterized by its genetic information, which is a complete set of variable values. By changing the genetic information, the population develops towards an optimum fit. To this end, the initial sample is partitioned into several sub-samples, namely complexes. In each complex, the varying combination of points produces offspring by the downhill simplex procedure of Nelder (1965). The probability of an individual taking part in the reproduction is proportional to its fitness. During the process, points with lower fitness are replaced by the offspring.

The SCE in dam network operation was coupled with DHMs to ingest the QPF during real-time measurement, as shown in Figure 11. More details on the coupling procedure can be found in Saavedra Valeriano et al. (2010a).

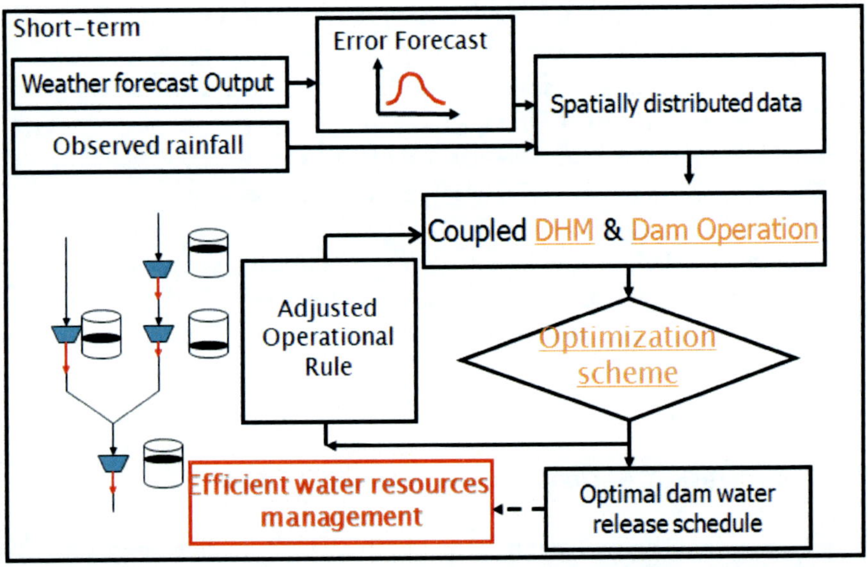

Figure 11. Coupling of a distributed hydrologic model and an optimization Scheme for dam network operation. The uncertainty of the forecast rainfall is to be reflected at the release schedule.

The objective is to minimize the flood peak at control points downstream and reduce the available free volume at reservoirs (which inevitably leads to an increase of water volume). A tradeoff is attempted between water allocation at the control point, which might result in flood damage, and water storage in the reservoirs. A proper objective function is defined to achieve optimal dam operation, in which flow release is the decision variable (Saavedra Valeriano et al., 2010a).

3. Flood Control Systems

This section presents two types of flood control systems for supporting dam operators and early flood warning emission. These systems take into account QPF's performance from previous time steps to support appropriate and timely decisions in near future.

3.1. Dam Release Support and Early Flood Warning Systems

A dam release support system (DRESS) was developed as a structural flood control example; it is summarized in Figure 12. The DBiHM was chosen as the hydrologic model, which runs in normal mode, forced by observed precipitation data, until a heavy precipitation event is detected. At that point, forecast error is evaluated. Recent observed precipitation and QPFs are compared over the previous time step window, in terms of location, intensity, and extent. Next, a weighting method calculates weights in each zone. These weights define the amplitude of QPF perturbation, generating ensemble precipitation members over the total lead time (Figure 12). Then, the DBiHM model is run to obtain the ensemble streamflow forecast. For special dam operations under heavy precipitation, a priori dam release is calculated for each ensemble member. Once the suggested optimal dam release schedule is determined using the QPF signal, the efficiency is evaluated using observed precipitation.

A nonstructural flood control measure is the early flood warning system (EFWS) in real-time, which uses a deterministic QPF introduced in Section 2.1.2. The system runs with precipitation forecasts (t0), to demonstrate real-time feasibility. Alternatively, the evaluation of precipitation forecasts for one previous time step (t−1) is possible, as recent observations become available. These spatial and temporal evaluations define the perturbation spread of the precipitation forecast (t0). The precipitation input data force the hydrologic model to generate ensemble river discharge, and then the probability of exceeding the flood warning level can be obtained. This information is expected to be more effective in supporting decision makers, because the associated uncertainty is provided as model output (Figure 13).

Both systems use QPF and real-time precipitation measurements as inputs. For cases where dams are operating in the basin, recent measurements should also be provided. After QPF from the previous time step is evaluated, a weighting system for actual QPF is attained. These weights define the perturbation amplitude for the ensemble precipitation. Then, the ensemble members force a DHM (e.g., GBHM or DBiHM) to generate reservoir inflows (for the DRESS), or river discharge at a control point (for the EFWS). Then, QPF inaccuracy can be reflected for dam operation or the flood warning level, as summarized in Figure 14.

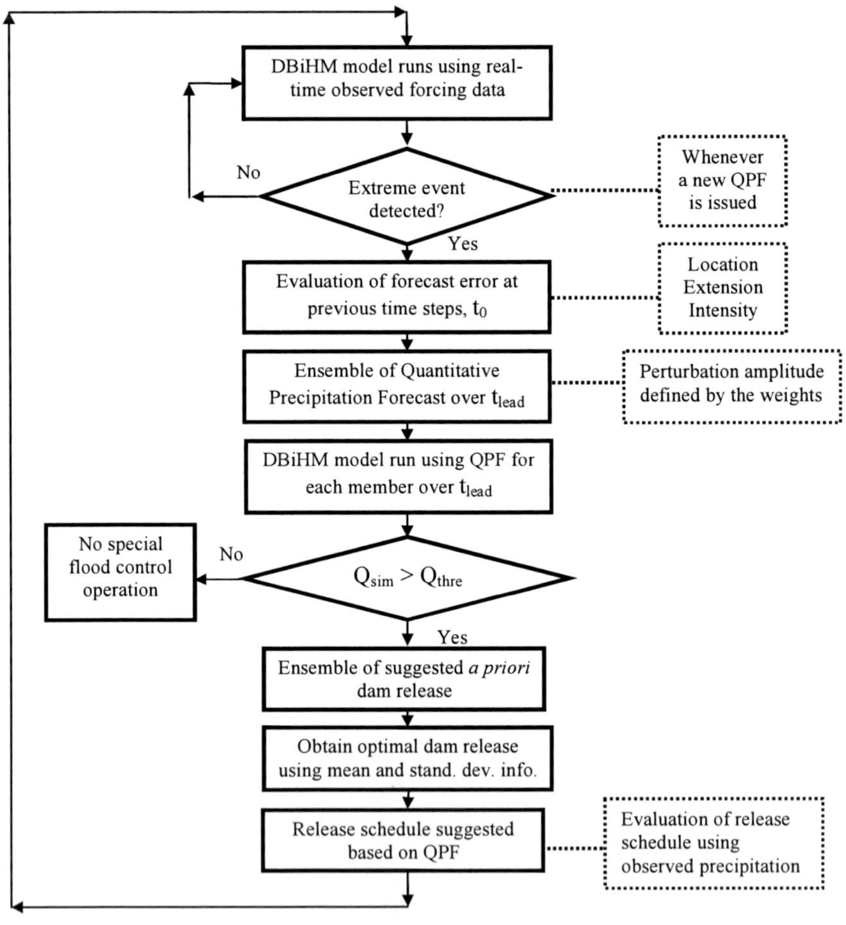

Figure 12. Steps of the dam decision support system DRESS for flood control under extreme events (Saavedra Valeriano et al., 2010b).

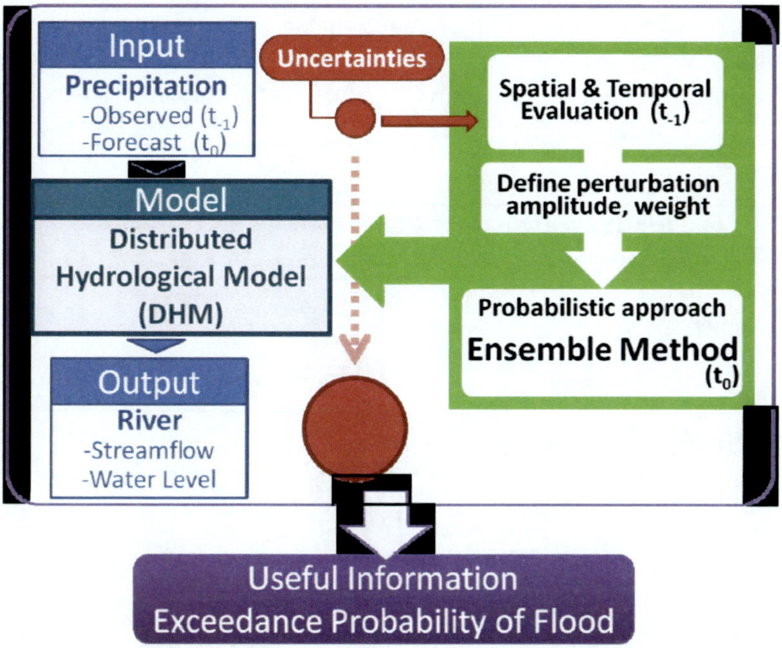

Figure 13. Overview of proposed Early Flood Warning (EFW) System. The system uses quantitative precipitation forecast(qpf) and real-time measurements. Recent time steps are evaluated, and the uncertainties are reflected as perturbation amplitude. The ensemble method suggest different members to force the DHM model. Finally, the performance of t simulated stream flow is evaluated as exceedance probability to support decision makers.

3.2. Case Study of Dam Release Support System for Upper Tone Reservoir System, Japan

The efficiency of DRESS is demonstrated by applying it to the Upper Tone reservoir system in Japan during the rainy season (June through September). Soil and vegetation associated parameters in WEB-DHM were calibrated and validated using data observed from 2000–2001 and 2002 respectively. First, simulated river discharge was compared with values recorded in situ. Second, the pattern of land surface temperature obtained from the model was compared against distributions acquired by satellite. The WEB-DHM had a 1 h time step, whereas the optimization horizon was 18 h, according to the QPF lead-time. Initial conditions were set to several hours prior to the event, based on observed data. To examine flood reduction at the

Maebashi gauging station, three multipurpose dams arranged in parallel, Fujiwara, Aimata, and Sonohara, were selected (Figure 15). Once an extreme event is identified, the system starts by verifying the situation at the control point, showing potential flood peaks.

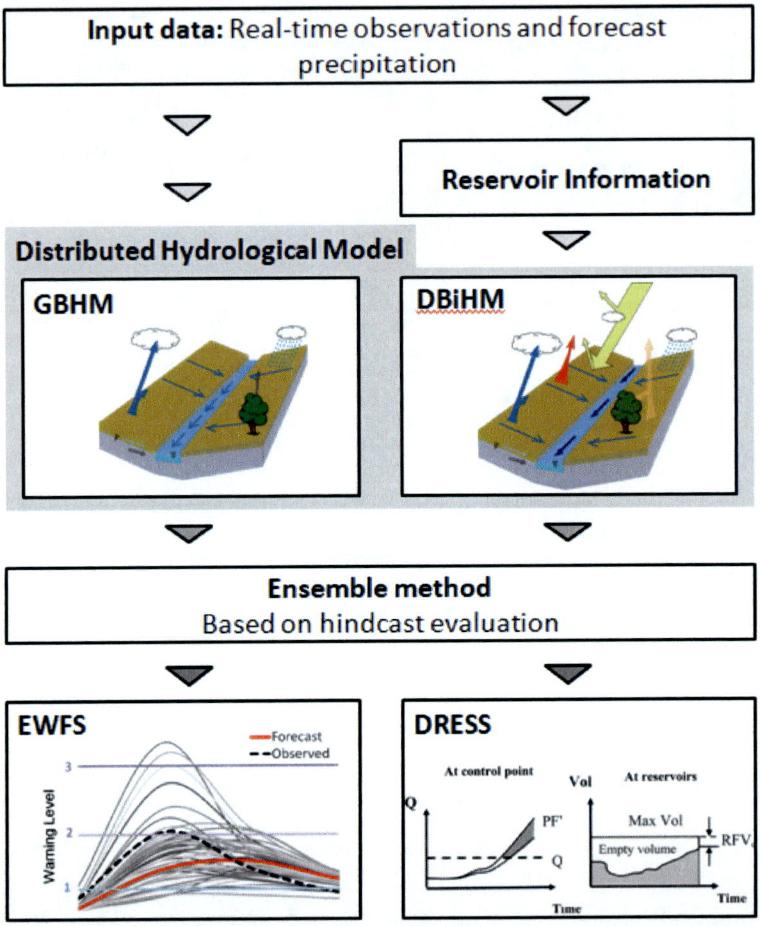

Figure 14. Summary of both decision support systems: DRESS and EFWS. Both require QPF and real-time measurements and the hydrological models to simulate the processes. DRESS required actual dam operation form recent time steps. The evaluation will decide the amplitude of perturbation of ensemble members. The uncertainties of the QPF are reflected in the dam release and warning level for DRESS and EFWS respectively.

The DRESS was applied under real-time operation, using recent 18-h data issued by meso-scale model-grid point value (MSM-GPV). Thus, the simulation is updated every 6-h in real-time operation.

The QPF applied to the Upper Tone catchment was obtained from the operational, non-hydrostatic Japan Meteorological Agency (JMA) model (Saito et al., 2007). This model has a 10-km horizontal resolution, and was issued four times a day through February 2006 (data are available at the University of Tokyo website, http://gpv.tkl.iis.u-tokyo.ac.jp/GPV/). The MSM-GPV dataset was downloaded and clipped onto the study area, which covers all of Japan (2900 km × 3600 km); the most relevant heavy rainfall events from 2002 to 2004 were selected. The dataset has a total lead-time up to 18 h, updated every 6 h. Thus, for each hour of observed precipitation, three different precipitation forecast values are available (i.e., 1–6, 7–12, and 13–18 h).

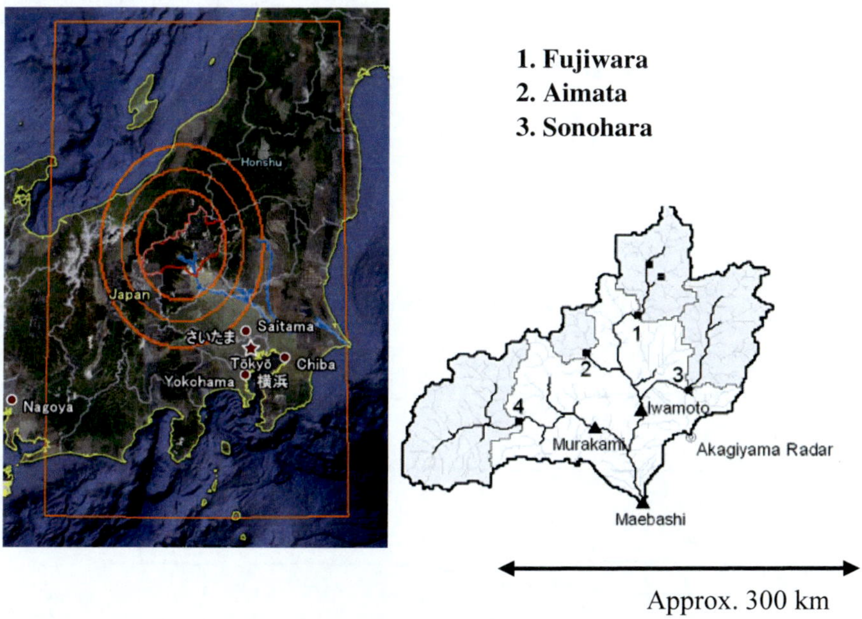

Figure 15. Location of upper Tone River system northwest of Tokyo, Japan encompasses 3304 km^2. The left figure shows the proposed zones around the basin to evaluate the forecast error. The right figure depicts the dams' arrangement in squares and the stream flow gauging stations in triangles.

Integrating the proposed dam release schedule over the evaluation time steps produced a release schedule per event. Then, the WEB-DHM model was run in evaluation mode, using observed precipitation from the Automated Meteorological Data Acquisition System. The flood reduction at Maebashi gauging station, using the proposed release schedule of the three dams for the 2002 event, is shown in Figure 16.

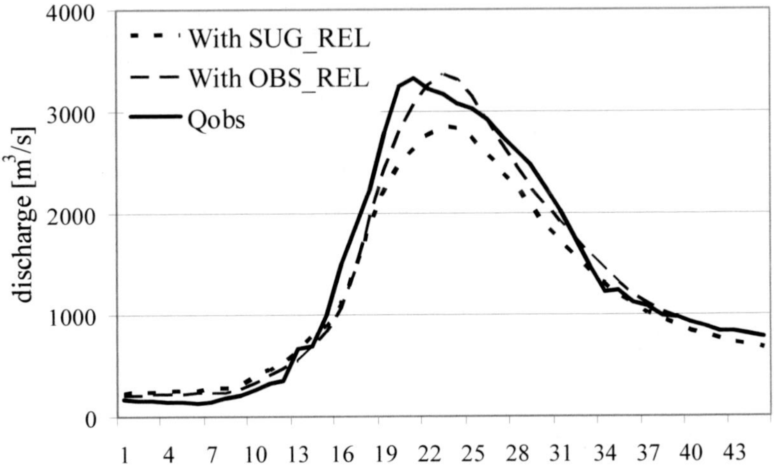

(a)

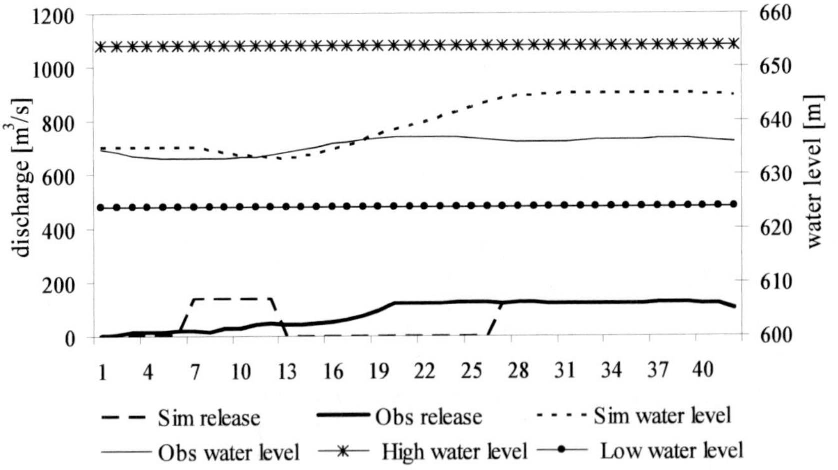

(b)

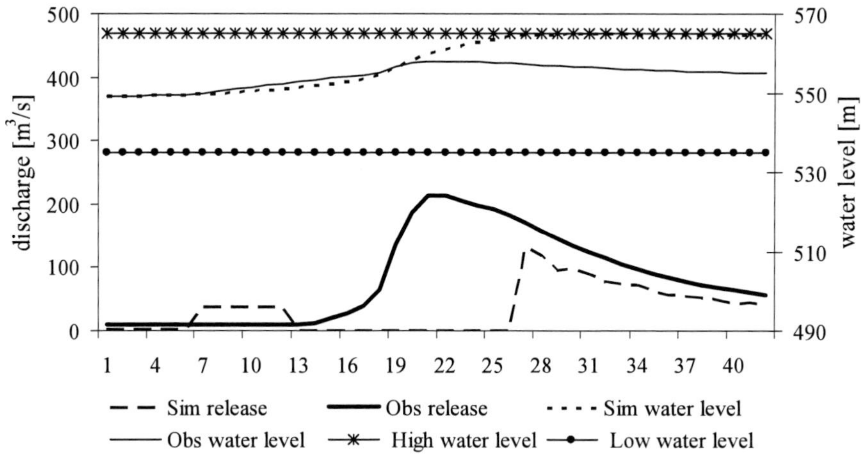

(c)

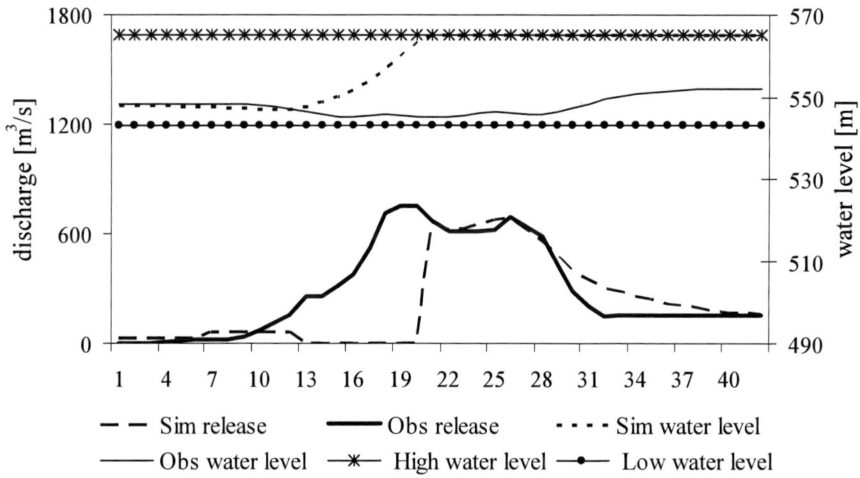

(d)

Figure 16. Flood reduction at Maebashi gauging station (**a**) using proposed release from DRESS system. At this case it was operated three dams namely: (b) Fujiwara, (c) Aimata, and (d) Sonohara. The system was forced by quantitative precipitation forecast with 18-h lead-time starting on 10 July, 2002. It can be seen that further flood peak reduction was possible at Maebashi (a) compared to the observed release since proper and timely flushing (b-d) was possible once the event was forecast. The upper lines of b, c, d show the water level fluctuation during the operation, at the end simulated water levels using suggested release rates allow replenish of volumes which could be used for water use and power generation.

The DRESS, which was developed for flood control during heavy rainfall, shows promising results in real-life operation. It uses the WEB-DHM model forced by the MSM-GPV, an operating mesoscale QPF in Japan. To reflect recent forecast errors in the near-future dam release schedule, an ensemble forecast approach and optimization algorithm were used. An objective function was formulated to reduce flood effects downstream and increase the total volume available in the reservoirs.

The DRESS was also validated by Saavedra Valeriano et al. (2010b), using the heaviest rainfall events of 2003 and 2004 in the Upper Tone River reservoir network. The efficiency of the system was demonstrated for all selected events, with some differences dependent on the accuracy of the precipitation forecast by MSM-GPV. The objective function successfully minimized the flood peak at Maebashi control point and the free volume in the reservoirs, by considering different release scenarios. The results indicate that DRESS is feasible for real-life dam operation.

3.3. Case Study of Early Flood Warning System at Huong River, Vietnam

The Huong River basin in Thua Thien Hue Province of central Vietnam (Figure 17) is affected by floods annually. Several of these events have resulted in human casualties and material losses in Hue City. In response to the need for reducing flood impacts, various national policies and operations have been implemented, such as:

(1) The Vietnamese government has started enforcing a strategy for disaster prevention and mitigation, which will continue to the year 2020.
(2) The Vietnam National Center for Hydro-Meteorological Forecasting (NCHMF) issues 6-h lead-time forecasts using the High Resolution Model (HRM), with both real-time observations at rain gauges and water levels at upstream stations.

Meanwhile, some related studies have also been conducted:

(1) Estimation of water amount arriving in flooded areas of the coastal plain, using the Stream model (Villegas, 2004).

(2) Improvements to flood hazard maps for the Huong River basin, by applying statistical analyses of inundation for Thua Thien Hue (NCAP, 2005).
(3) Flood simulation using a DHM with various precipitation input data (Saavedra Valeriano et al., 2009). To further improve this approach for implementation, governmental operations should cooperate with researchers. The Tokyo Institute of Technology, University of Tokyo, and Vietnam NCHMF are currently collaborating on flood forecasting in the Huong River basin, which is a demonstration basin within the Global Earth Observation System of Systems and the Asian Water Cycle Initiative.

Since the regional forecast HRM is not systematically archived for the Huong River basin, a global QPF is hereby proposed as a feasible alternative. The global-scale model (GSM) output provided by the JMA was used. This dataset provided global QPF, i.e., grid-point values (GPV) at 1.25° spatial resolution, issued every 12 h and recorded every 6 h within the total lead time (up to 24 h in this study). The dataset has been archived from July 2002 to the present. The GSM-GPV data are archived at the University of Tokyo. Because the GPV forecast is only at one point in the catchment area, this point value was applied to the entire basin uniformly. Hence, the GPV forecast is always available with two scenarios each time step, because of lead-time overlap. Therefore, it is proposed to divide the GPV forecast into two parts, and evaluate each tendency; 0–12 h lead-time is called GPV12, and 13–24 h lead-time is called GPV24.

Satellite-based precipitation was procured from the Tropical Rainfall Measuring Mission (TRMM), which is a joint mission between the National Aeronautics and Space Administration (NASA) and the Japan Aerospace Exploration Agency. TRMM provided a calibration-based sequential scheme for combining precipitation estimates from multiple satellites and gauge analyses. This combination is done where feasible, at fine scales (0.25° × 0.25°, every 3 h). The dataset covers the latitude band 50°N–50°S, from 1998 to the present (Huffman et al., 2007). Since both input rainfall values are point data, they were spatially integrated using Thiessen polygons, to obtain values at each simulation grid cell for the hydrologic GBHM model.

The application targeted the rainy seasons (September through November) of 2004 and 2006; both periods had normal floods and inundation, caused by extreme flood peaks. The model forced by global QPF simulated the discharge amount during the targeted periods. The time step of the hydrologic model was 1 h; hence, temporal downscaling was done from 6-h precipitation data.

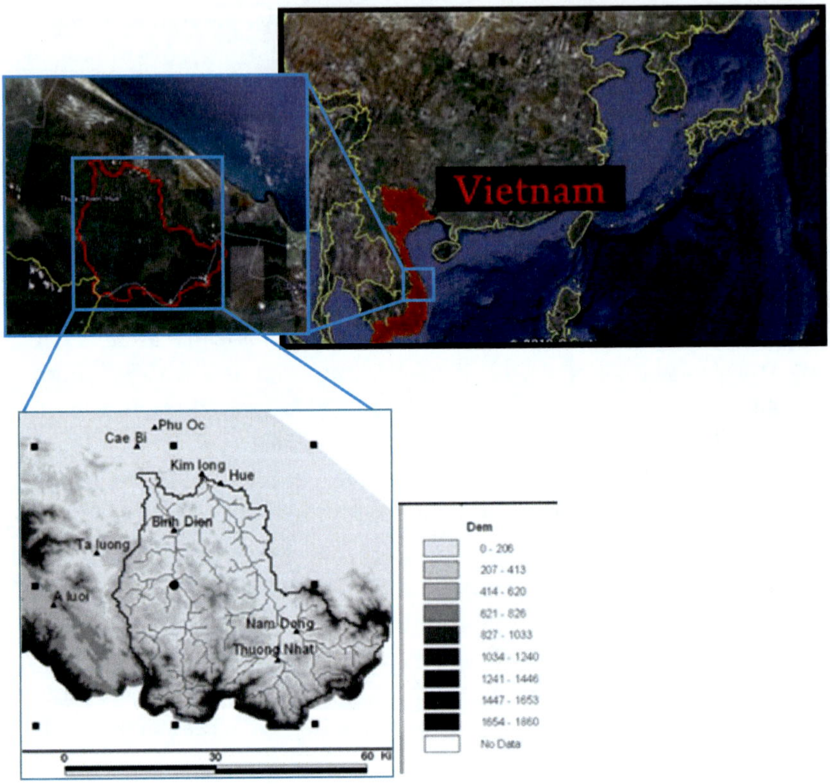

Figure 17. Watershed delineation on the Huong River basin, in Hue, Vietnam.

At least seven floods of warning level 2 (corresponding to a water level of 2.0 m) occur every year at Kim Long Station, Hue. On average, the river flood level remains at warning level 2 for a few days; for some events, this can be up to 9 days. The flood level variation every year is relatively large. According to data from 1977 to 2005, there was an annual average of 3.5 floods of threat magnitude equal to or greater than warning level 2 on the Huong River. The flood season begins in October and continues through the first half of December. This is the period of strong typhoons and tropical cyclones, which produce heavy rain over a large area. The floods during this season often have high, double peaks (NCAP, 2005).

Table 4. Prediction Performance of Flood warning level suggested by EFW system based on GPV12

Date	Prediction exceedance probability at each water level				Observed water level (m)
	0m	1m	2m	3m	
2-Oct	100	0	0	0	0.8
3-Oct	100	0	**0**	0	2.0
4-Oct	100	**0**	0	0	1.4
5-Oct	100	0	0	0	0.9
27-Oct	100	0	0	0	0.6
28-Oct	100	**0**	0	0	1.1
29-Oct	100	0	0	0	0.9
30-Oct	100	0	0	0	0.7
16-Nov	100	0	0	0	0.5
17-Nov	100	**0**	0	0	1.0
18-Nov	100	0	0	0	0.8
24-Nov	100	0	0	0	0.9
25-Nov	78	22	0	**0**	3.1
26-Nov	52	48	0	**0**	3.5
27-Nov	52	48	0	**0**	3.5
28-Nov	82	18	**0**	0	2.4
29-Nov	100	**0**	0	0	1.7
30-Nov	100	**0**	0	0	1.2

It is compared to actual level observation at most right column. At Kim Long station the flood warning level 1 corresponds to 1 m of water level. Similarly situation happens for 2 and 3 levels. The latter represents the most dangerous situation suggestion evacuation.

The hydrologic GBHM model was driven by each type of precipitation input data, to simulate streamflow at the Kim Long control point (Figure 17). The simulated periods were from 2 October to 30 November 2004, and from 23 September to 5 October 2006. Flood peaks exceeding 1 m water level occurred four times in 2004, and twice in 2006. The model was able to simulate extreme flooding using both raingauge and TRMM observations (24–30 November 2004, and 30 September to 4 October 2006), with some differences in the timing of the peak and the amount of discharge (Figure 18). In general, simulated streamflow using raingauge data performed better than that using TRMM data during normal flood events. The RMSE for raingauge and TRMM simulations was 133.2 and 234.8 in 2004 and 245.9 and 539.4 in 2006, respectively. In fact, it seems that TRMM data cannot precisely measure small amounts of precipitation; therefore, TRMM almost missed small flood events. For an extremely large flood in 2004, raingauge and TRMM appeared to be in better agreement, but both overestimated observed discharge. This tendency is acceptable, however, because the actual stream overflowed and river discharge could not be measured (Saavedra Valeriano et al., 2009). Both GPV forecast series tended to underestimate flood magnitude except for the extreme flood of 2006, for which there was overestimation. The GPV12 tends to underestimate compared with GPV24, as in the events of 2004. This might be caused by the spin-up time of the NWP required to reach a stable state. The GPV24 showed better performance in 2004, even for normal flood events.

The TRMM performance appeared to be sufficient for simulating streamflow, although it was less precise than the raingauge method during the target years. GPV24 showed its best performance in 2004, and it was poorer in 2006. In fact, GPV24 overestimated the precipitation amount during the extreme flood of 2006. GPV12 performed better for that event.

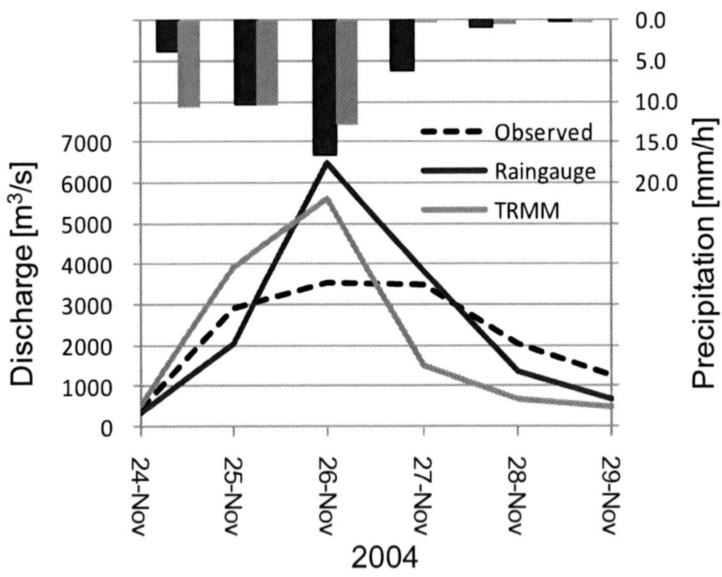

(a)

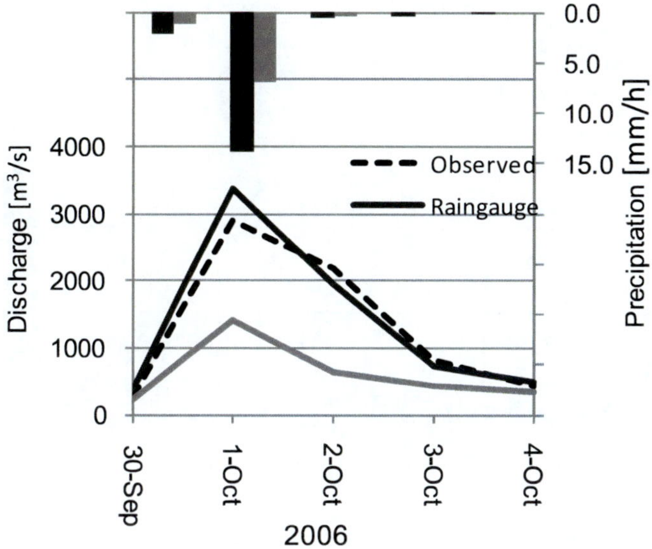

(b)

Figure 18. Simulations of discharge using raingauge and TRMM data at Kim Long station in comparison with observed discharge in (a) 2004 and (b) 2006.

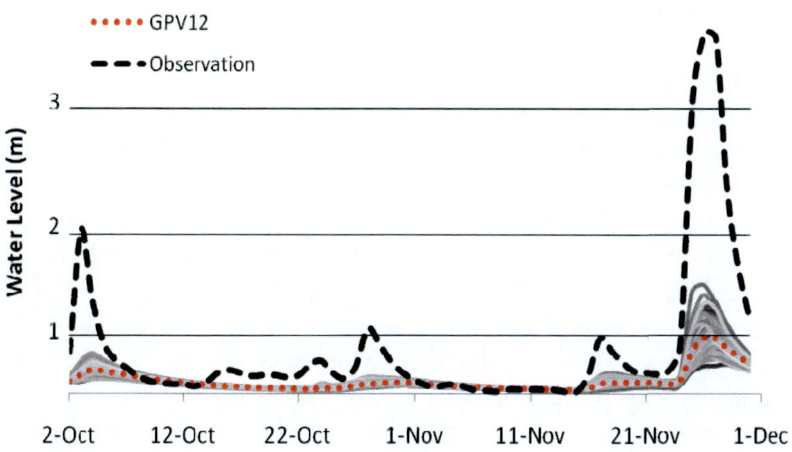

Figure 19. Simulation of water level at Kim Long station. Fifty ensemble members in gray lines were generated using Global Quantitative Precipitation forecast with 12 h lead-time (i.e. GPV12). Even though underestimation is dominant, at end of Nov. 2004 the ensemble members suggest higher magnitude of water level.

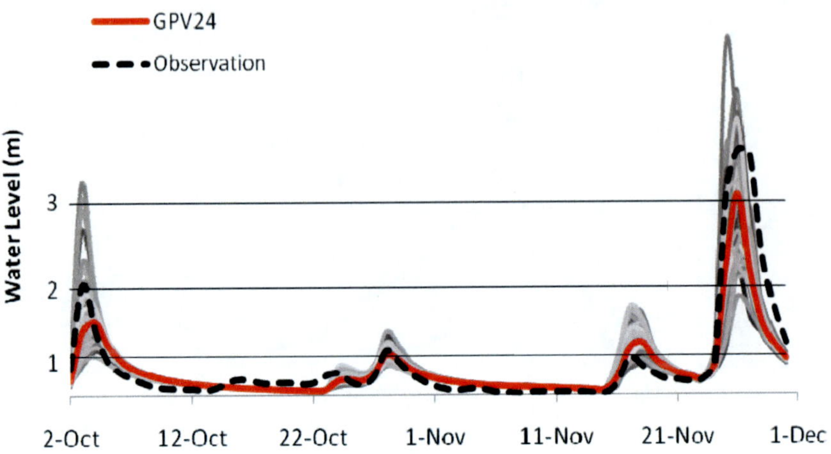

Figure 20. Simulation of water level at Kim Long station. Fifty ensemble members in gray lines were generated using Global Quantitative Precipitation forecast with 24 h lead-time (i.e. GPV24). GPV24 shows much better agreement with the observed than that of GPV12.

The EFWS based on the real-time ensemble method was developed, using global QPF and quasi-global, satellite-based precipitation data. Even with a GPV forecast at coarse resolution, the EFWS was used successfully in the Huong River basin. In fact, it was found that the GPV forecasts the timing of heavy rainfall with acceptable accuracy, as compared with both raingauge and TRMM measurements. However, the magnitude was not predicted accurately. The 24-h lead-time global forecast is crucial for preparedness and reduction of casualties and material losses.

Using the ensemble method based on the GPV bias evaluation to account for uncertainties, the EFWS predicted the timing and peak of flood events with some accuracy. The EFWS generated sufficient precipitation forecast members using previous time steps, enabling proper timing of flood warning levels. It was also confirmed that the performance of satellite-based TRMM measurements may be adequate for humid, vegetated basins. In this study, TRMM alone performed suitably in the Huong River basin during the evaluation rainy seasons. This may not necessarily be the case for other River basins, however. It is therefore advisable to verify the satellite-based precipitation products, using available raingauge measurements within or near a basin.

Table 5. Prediction Performance of Flood warning level suggested by EFW system based on GPV24

Date	Prediction exceedance probability at each water level				Observed water level (m)
	0m	1m	2m	3m	
2-Oct	98	2	0	0	0.8
3-Oct	10	76	**10**	4	2.0
4-Oct	2	**98**	0	0	1.4
5-Oct	6	94	0	0	0.9
27-Oct	100	0	0	0	0.6
28-Oct	48	**52**	0	0	1.1
29-Oct	74	26	0	0	0.9
30-Oct	100	0	0	0	0.7
16-Nov	98	2	0	0	0.5
17-Nov	22	**78**	0	0	1.0
18-Nov	8	92	0	0	0.8
24-Nov	94	6	0	0	0.9
25-Nov	0	28	58	**14**	3.1
26-Nov	0	4	42	**54**	3.5
27-Nov	0	24	76	**0**	3.5
28-Nov	0	100	**0**	0	2.4
29-Nov	0	**100**	0	0	1.7
30-Nov	92	**8**	0	0	1.2

It is compared to actual level observation at most right column. At Kim Long station the flood warning level 1 corresponds to 1 m of water level. Similarly situation happens for 2 and 3 levels. The latter represents the most dangerous situation suggestion evacuation.

To improve the developed EFWS and prepare for real-life operation, it should be tested under real-time conditions, in parallel with the current system at NCHMF in Vietnam. In addition, forecasts for areas surrounding the basin can be useful in evaluating uncertainty, when the forecast is geographically misplaced (Saavedra Valeriano et al., 2010b). The current study suggests that this basic analysis of ensemble streamflow forecasts can aid decision makers in real time. Since GPV is a global forecast, it can be used in other basins of the world under similar conditions—heavily vegetated, steep, humid areas—to support the reduction of flood damage.

CONCLUSION AND THE FUTURE

Flooding as a societal water issue was addressed, including the current situation and future expectations from potential climate change. Advancements in operational QPF were then summarized. It is currently possible to predict the weather several days in advance with acceptable reliability, and there is also greater understanding of the physical factors affecting the lower atmospheric boundary layer. The spatial patterns of QPF can be easily assimilated by physically-based DHM to simulate streamflow within a river network. These types of hydrologic models have been evolving rapidly over the last few decades, owing to reduced computing cost and greater high-resolution data availability from remote sensing and real-time ground measurements.

Two DHM models were briefly introduced: (1) The GBHM, which has a fast computing capability and relies on flow intervals within sub-basins; and (2) the WEB-DHM, which includes explicit soil-vegetation-atmosphere interactions. According to the weather characteristics and setting of a basin, GBHM or WEB-DHM may be used to express the spatial variability of basin hydrologic processes. For example, for basins where potential evaporation exceeds precipitation, WEB-DHM can more accurately simulate soil moisture. Only after the hydrologic models are calibrated and validated with historical records, they can be used with QPF for real-time flood forecasting, extending lead-time.

However, QPF must be evaluated, since it can be overestimated or underestimated. Therefore, mathematical expressions were presented to define QPF forecast error, comparing with recent real-time precipitation measurements. Once an appropriate error expression is selected, the intensity and displacement of the main precipitation pattern is evaluated, using buffer zones around the basin. With this method, even precipitation events outside the watershed, captured by recent QPF issuances, can be reflected in the basin flood countermeasures. New techniques like data assimilation may be used for this purpose. After defining appropriate spread, the ensemble method based on noise perturbation generates ensemble precipitation members. With these members, the hydrologic model can generate an ensemble streamflow. This technique allows QPF uncertainty within the streamflow or stage forecast to be transmitted as valuable decision support information.

Two types of flood control approaches were reviewed: structural and nonstructural measures. In line with these two approaches, it is appropriate to consider two systems to support dam operation and early warning systems.

The systems are: (1) the DRESS; and (2) the EFWS. Both systems showed the capability to assimilate and reflect the QPF forecast from previous time steps within the streamflow forecast. The DRESS focuses on uncertainty in the real-time dam release schedule, whereas the EFWS targets real-time flood warning level forcing of DHMs by ensemble QPF. Probability analysis of the ensemble streamflow facilitates useful information for decision makers. The DRESS was validated within the operational river management of the Tone Reservoir system in Japan, using fine-resolution QPF and real-time precipitation data to optimize the dam release schedule. The efficiency of DRESS has been demonstrated for heavy rainfall events, for which the accuracy of mesoscale QPF was reflected in the releases from three dams, with 18 h lead-time. The objective function of the optimization algorithm was set to minimize the flood peak at control points at downstream dams, and to increase volume in the reservoirs. The results indicate that it is possible in real-life dam operation to decide on release from reservoirs, using QPF through the optimization tool. Similarly, the EFWS was validated using global and regional QPFs at Huong River basin in central Vietnam, which had few available rain gauges. The system succeeded in predicting the timing and peak of flood events with acceptable accuracy at a control point. This point is used as a reference to define and transmit the flood warning level to Hue city. Two rainfall events were used with a 24 h lead-time, which would have provided enough time for evacuation. It was also confirmed that, in combination with available in situ gauges, the performance of satellite-based precipitation measurements was adequate. The EFWS showed promising results, suggesting that it can be used in other developing countries, even those with limited observation networks.

Even though QPF has attained a sufficient reliability for use in real-life situations, the status of real-time flood forecasting using QPF has not improved recently. The advantage of QPF in operations would be to extend lead time, which is critically important for structural and nonstructural flood control. Unfortunately, as lead time is extended by several hours, accuracy may be sacrificed. These constraints should be overcome in the future by improving the prediction of numerical models. Development of correlations between QPF accuracy and the type of rainfall event (e.g., frontal, convective, typhoon), including physical understanding, would be a desirable parallel effort. In addition, it is important to express the output of models using QPF in a format easily comprehensible to decision makers.

As their economies grow in southeast Asian countries, there is greater investment in monitoring systems and structural flood control. Consequently, river basins like that of Huong River are moving toward an early warning system combined with structural flood control. Thus, the optimization of dam operation with QPF is important. The objective function should include other dam operation purposes besides flood control, such as water use, environmental flow, and hydropower. Regional forecasting capabilities at meteorological agencies have improved in recent years, leading to better understanding and prediction of the factors causing heavy rainfall.

Finally, as its accuracy improves on global to local scales, QPF can be incorporated in decision support systems. This will help societies be better prepared for catastrophic events like typhoons, thereby reducing flood damage.

References

Abbott, M.B., J.C. Bathurst, J.A. Cunge, P.E. O'Connel, and J. Rasmussen, (1986), An Introduction to the European Hydrological Systems-Systeme Hydrologique Europeen, 'SHE'. 2. Structure of a Physically Based Distributed Modelling System. *Journal of Hydrology* 87, pp. 61-77.

Berz, G. (2000), "Flood Disasters: Lessons from the Past - Worries for the Future." Proceedings of the Institution of Civil Engineers - *Civil Engineering*, 142, Issue I, March, pp. 3-8.

Beven, K J and M. J. Kirkby (1979), A physically based variable contributing area model of basin hydrology, *Hydrol. Sci. Bull.*, 24(1), pp 43-69.

Buizza, R. (2008), The value of probabilistic prediction, *Atmospheric science letters*, 9, pp. 36-42.

Carey, W.P., and A. Simon (1984), *Physical basis and potential estimation techniques for soil erosion parameters in the Precipitation-Runoff Modeling System* (PRMS): U.S. Geological Survey Water-Resources Investigations Report 82-4218, pp 32.

Collischonn, W., R. Hass, I. Andreolli, and C.E. Morelli Tucci, (2005), Forecasting River Uruguay flow using rainfall forecasts from a regional weather-prediction model, *Journal of Hydrology*, 305, pp. 87-98.

Chang, L. C. and F. J. Chang (2009), Multi-objective evolutionary algorithm for operating parallel reservoir system, *Journal of Hydrology*, 377, pp. 12-20.

Chen, L. (2003), Real coded genetic algorithm optimization of long term reservoir operation. *Journal of American Water Resources Association*, 39(5), pp. 1157-1165.

Cloke, H.L. and Pappenberger, F. (2009), Ensemble flood forecasting: A review, *Journal of Hydrology*, 375, pp. 613-626.

Duan, Q., S. Sorooshian, and V.K. Gupta (1992), Effective and Efficient Global Optimization for Conceptual Rainfall-Runoff Models, *Water Resources Research*, 28(4), pp. 1015-1031.

Ebert, E.E., and L.L. McBride (2000), Verification of precipitation in weather systems: determination of systematic errors, *Journal of Hydrology*, 239, pp. 179-202.

Golding, B.W. (2000), Quantitative precipitation forecasting in the UK, *Journal of Hydrology*, 239, pp. 286-305.

Hirabayashi, Y. , Kanae S., Emori, S., Oki, T. and Kimoto, M. (2008), Global projections of changing risks of floods and droughts in a changing climate, *Hydrological Sciences Journal*, **53**(4), pp. 754-772.

Honda, Y., M. Nishijima, K. Koizumi, Y. Ohta, K. Tamiya, T. Kawabata, and T. Tsuyuki (2005), A pre-operational variational data assimilation system for a non-hydrostatic model at the Japan Meteorological Agency: Formulation and preliminary results, *Quarterly Journal of the Royal Meteorological Society*, 131, pp. 3465-3475.

Huffman, G. J., Adler, R. F., Bolvin, D. T., Gu, G. J., Nelkin, E. J., Bowman, K. P., Hong, Y., Stocker, E. F., and Wolff, D. B. (2007), The TRMM multisatellite precipitation analysis (TMPA): Quasi-global, multiyear, combined-sensor precipitation estimates at fine scales, *Journal of Hydrometeorology*, 8, pp. 38-55.

Netherlands Climate Assistance Programme (2005), Climate change impacts in Huong River basin and adaptation in its coastal district Phu Vang. Project Report, Hanoi, Vietnam.

Jasper K., J. Gurtz, and H. Lang (2002), Advanced flood forecasting in Alpine watersheds by coupling meteorological observations and forecasts with a distributed hydrological model, *Journal of Hydrology*, 267, pp. 40-52.

Krzysztofowicz, R., and C. G. Collier (2004), Quantitative precipitation forecasting II, *Journal of Hydrology*, 288, Issues 1-2, p. 1

Labadie, J. W. (2004), Optimal Operation of Multireservoir Systems State-of-the-Art Review, *Journal of Water Resources Planning and Management, ASCE*, 130(2), pp. 93-111.

Lynch, P. (2008), The origins of computer weather prediction and climate modeling, *Journal of Computational Physics,* 227, pp. 3431-3444

Madsen, H. and T. Jacobsen (2001), Automatic Calibration of the MIKE SHE Integrated Hydrological Modelling System. *4th DHI Software Conference, Helsingør, Denmark, 6-8 June.*

Mays, Larry W. (2005), Water resources Engineering, , *John Wiley & Sons, Inc*. pp.534-556

Nelder, J.A. and Mead, R. (1965), A simplex method for function minimization, *Computer Journal*, 7, pp. 308–313.

NetCatSERVICE (2011), http://www.munichre.com/en/reinsurance/business/non-life/georisks/ natcatservice/ default.aspx

Palmer, T. and Buizza, R. (2007), Fifteenth anniversary of EPS. ECMWF Newsletter 114 (Winter 07/08), 14.

Saavedra Valeriano, O. C., T. Koike, K. Yang, D. Yang, C. T. Nyunt, D. V. Khanh, and L.C. Chau (2009), Flood simulation using different sources of rainfall in the Huong River, Vietnam. *Hydrological Sciences Journal*, 54(5), pp. 909-917.

Saavedra Valeriano, O. C., T. Koike, K. Yang, and D. Yang (2010a), Optimal Dam Operation during Flood Season using a Distributed Hydrological Model and a Heuristic Algorithm, *Journal of Hydrologic Engineering*, 15(7), pp. 580-586.

Saavedra Valeriano, O. C., Koike, T., Yang, K., Graf, T., Li, X., Wang, L. & Han, X. (2010b): Decision support for dam release during floods using a distributed biosphere hydrological model driven by quantitative precipitation forecasts, *Water Resources Research*, 46, W10544, doi:10.1029/2010WR009502, pp. 1-13.

Saito, K., J.I. Ishida, K. Aranami, T. Hara, T. Segawa, M. Narita, Y. Honda (2007), Nonhydrostatic atmospheric models and operational development at JMA, *Journal of The Meteorological Society of Japan*, 85(B), pp. 271-304.

Sellers, P. J., D. A. Randall, G. J. Collatz, J. A. Berry, C. B. Field, D. A. Dazlich, C. Zhang, G. D. Collelo, and L. Bounoua (1996), A revised land surface parameterization (SiB2) for atmospheric GCMs, Part I: Model formulation, *Journal of Climate*, 9, pp. 676-705.

Singapore red cross society (2010), The Pakistan Flood Update, 12 Oct 2010, http://www.redcross.org.sg/

Team, C. W., Pachauri, R. K., and Reisinger, A. (2007), Contribution of Working Groups I, II and III to the Fourth Assessment Report of the Intergovernmental Panel on Climate Change.

Yang, D., S. Herath, T. Oki., and K. Musiake (2001), Application of a distributed hydrological model in the Asian Monsoon Tropic Region with a Perspective of Coupling with Atmospheric Models, *Journal of the Meteorological Society of Japan*, 79(1B), pp. 373-385.

Yang, D., S. Herath, and K. Musiake, (2002), Hillslope-based hydrological model using catchment area and width functions, *Hydrological Sciences Journal*, 47, pp. 49-65.

Wang, L., T. Koike, K. Yang, T.K. Jackson, R. Bindlish and D. Yang (2009), Development of a distributed biosphere hydrological model and its evaluation with the Southern Great Plains Experiments (SGP97 and SGP99), *Journal of Geophysical Research*, 114, D08107.

Zhu, Y. (2005), Ensemble forecast: A New Approach to Uncertainty and Predictability, *Advances in Atmospheric Sciences*, 22(6), pp. 781-788.

INDEX

A

B

C

D

E

F

G

T

U

V

W

X

Y